T0276472

Handbook of Centrifugal Pumps

Handbook of Centrifugal Pumps

Edited by **Vincent Woods**

CLANRYE
INTERNATIONAL

New Jersey

Published by Clanrye International,
55 Van Reypen Street,
Jersey City, NJ 07306, USA
www.clanryeinternational.com

Handbook of Centrifugal Pumps
Edited by Vincent Woods

International Standard Book Number: 978-1-63240-257-8 (Hardback)

Printed in the United States of America.

Contents

Preface

This book aims to highlight the current researches and provides a platform to further the scope of innovations in this area. This book is a product of the combined efforts of many researchers and scientists, after going through thorough studies and analysis from different parts of the world. The objective of this book is to provide the readers with the latest information of the field.

Centrifugal pumps are commonly used for pumping sewage, petrochemicals etc. The structure of a hydraulic machine, as a centrifugal pump, has evolved primarily to fulfill the requirements of fluid flow. This book compiles various case studies which take into consideration the strong collaboration between the pump and the pumping installation, and the need to manage the process. It also discusses the necessity to operate best efficiency in order to save energy, the condition to advance the operation against cavitation and other significant topics. The book will be beneficial for readers interested in gathering information about centrifugal pumps.

I would like to express my sincere thanks to the authors for their dedicated efforts in the completion of this book. I acknowledge the efforts of the publisher for providing constant support. Lastly, I would like to thank my family for their support in all academic endeavors.

Editor

Analysis of Cavitation Performance of Inducers

Xiaomei Guo, Zuchao Zhu, Baoling Cui and Yi Li
The Laboratory of Fluid Transmission and Application,
Zhejiang Science Technology University,
China

1. Introduction

Low specific speed centrifugal pumps have low flow rates and high heads. They are widely applied in the petroleum, chemical, aerospace, pharmaceuticals, metallurgy, and light industries, among others. With the development of space technology and petrol chemical industry, the highly stable cavitation performance of centrifugal pumps has been put forward. Poor cavitation performance is one of the key problems in low specific speed centrifugal pumps. The most effective method for solving this problem is adding an inducer upstream of the impeller to identify the influence produced by different pre-positioned structures. This chapter focuses primarily on the analysis of the cavitation performance of inducers to identify the influence imposed by different inducers on the cavitation performance of a centrifugal pump. The chapter is organized into five sections. First, the status of research on cavitation performance is reviewed. Second, the research model is described. Third, the simulations of the different inducers are presented. Fourth, the cavitation performance experiment is carried out. The conclusion ends the chapter.

2. Research status

Numerical calculation techniques have developed rapidly in recent years, and many works have been carried out on inducer flow and its cavitation performance. The results of the one-phase simulation of single and serial inducers (Cui et al., 2006) show that inducers can increase impeller inlet pressure, the easy to cavitate position is located at the rim of the suction surface near the inlet, and cavitation does not take place in the second inducer. The flow in the screw inducer is numerically calculated (Wang Jian-ying & Wang Pei-dong, 2006), and the results show that the head can be efficiently increased by adding a screw inducer. Guo et al. (2010) carried out a simulation of the flow in two different inducer structures, and showed that parameters including helical pitch, axial length, and blade wrap angle pose considerable influence on cavitation. Cavitation is an important phenomenon in the design of an inducer. The understanding and prediction of the mechanisms associated with cavitation have progressed significantly the past few years. Unsteady flow in the equal pitch inducer is numerically calculated by adopting the cavitation and mixture model (Ding & Liang, 2009). The results show that the area prone to cavitation is the rim of the suction surface. Unsteady flow in the progressive pitch inducer is also calculated using the Euler multiphase model and standard k-ε turbulence model (Yuan et al., 2008; Kong et al., 2010). The findings show that rounding out the blade inlet can improve the cavitation performance of the inducer. The rotation cavitation of one channel and four channels of the inducer is simulated by adopting the unsteady cavitation model (Langthjem & Olhoff, 2004). The complex cavitation flow of the inducer is solved using the CRUNCH program and multiple unstructured grids (Li, 2004). The inducer's cavitation performance is determined through the simulation (OkitaK et al., 2009; Li & Wang, 2009). In references (Kunz et al., 2000; Medvitz et al., 2001), a preconditioned Navier-Stokes method has been applied to calculate cavitation flows in centrifugal pumps. Some numerical works have been developed to predict cavitation inception, cavity dimensions, and/or thresholds corresponding to pump head drops (Ait-Bouziad et al., 2003, 2004; Mejri et al., 2006). Researchers (Hosangadi & Ahuja, 2001) used a hybrid unstructured mesh to simulate the cavitation flows over a hydrofoil and a cylindrical headform. Hosangadi (2006) presented a good comparison of simulated and experimental data on breaking down a helical flat-plate inducer configuration in cold water. The influence of steady cavitation behavior on pump characteristics and on the final head drops was also simulated (Benoît et al., 2008).

In spite of these relevant works, more studies are needed to improve on earlier achievements. To reveal the mechanism of two-phase flow in an inducer under cavitation conditions, four different inducers are designed, gas-liquid two-phase flows are simulated, and a corresponding external cavitation experiment is carried out. In this paper, the mixture model and standard k-ε turbulent model are adopted for the simulation. The inducer, impeller, and volute are made as an entire channel for simulation by adopting a gas-liquid two-phase model. During the simulation, the radial gap between the inducer blade tip is taken into account, and the value is 1 mm.

3. Research model

The research object is a high-speed centrifugal pump with an inducer (four different structures) upstream of the impeller (see in Fig.1). The flow rate is 5 m³/h, head 100 m, rotation speed 6 000 r/min. Seen from the inlet, rotation direction of inducer is clockwise. The centrifugal pump's impeller is shown in Fig.2. Four different inducers are adopted. One is equal-pitch. Second is long equal-pitch (with longer pitch than the first one). Third is progressive pitch. Fourth is with short splitting blades that with two long and two short blades (we call it two-long and two-short inducer in this chapter). The first three inducers are shown in Fig.3. Their parameters are shown in table 1. The last one is shown in Fig.4. Main geometry parameters are shown in table 2.

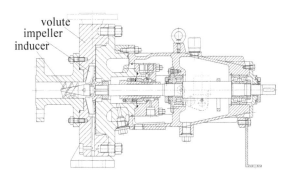

Fig. 1. The high-speed centrifugal pump with an inducer upstream of the impeller

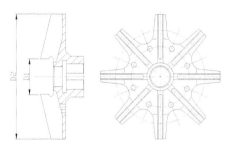

Fig. 2. Impeller

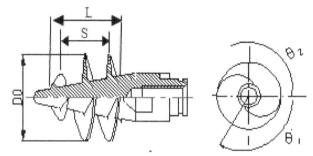

Fig. 3. Inducers with two blades

parameters	S	L	θ_1	θ_2
Equal-pitch inducer	20	24.3	120	317.5
Long-equal-pitch inducer	28	37	120	355.3
Progressive pitch inducer	$155.5\tan(\beta_i)$ $6.6 \leq \beta_i \leq 12.6$	33	120	354.3

Table 1. Parameters of the inducers with two blades

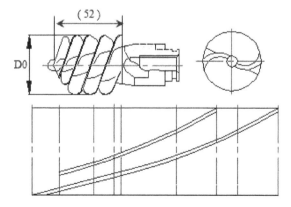

Fig. 4. Inducer with two long and two short blades

Geometry parameter	D0 (mm)	D1 (mm)	D2 (mm)
Diameter	38	28	136

Table 2. Main geometry parameters

4. Numerical simulations

4.1 Control equation

Mixture model is a simplified multiphase flow model, which is used to simulate different phases with different velocities. In many cases, the mixture model is a good alternative to Eulerian model, and it can get good results as other good multiphase model. The fluent software numerical code solves the standard k-ε turbulent model equations of a homogeneous fluid (Fortes et al., 2007, Coutier et al., 2004). Previous studies (Yuan et al., 2008, Tang et al., 2008, Benoît et al.,2008) pointed out that the Mixture model can successfully yield quantitative predictions of cavitation flow global parameters (i.e., characteristic frequencies, vapor structure size). As the gas-phase volume is relatively few when the inducer cavitates, the gas and liquid phases are supposed to be incompressible . So the mass equations are adopted as bellow:

$$\frac{\partial \rho_m}{\partial t} + \nabla \bullet (\rho_m \vec{v}_m) = \dot{m} \tag{1}$$

$$\vec{v}_m = \frac{a_1 \rho_1 \vec{v}_1 + a_2 \rho_2 \vec{v}_2}{\rho_m} \tag{2}$$

Where

$$\rho_m = a_1 \rho_1 + a_2 \rho_2 \tag{3}$$

$$\vec{v}_m = (a_1 \rho_1 \vec{v}_1 + a_2 \rho_2 \vec{v}_2) / \rho_m \tag{4}$$

Momentum Equation for the Mixture:

$$\frac{\partial (\rho_m \vec{v}_m)}{\partial t} + \nabla \bullet (\rho_m \vec{v}_m \vec{v}_m) = -\nabla p + \nabla \cdot [\mu_m (\nabla \vec{v}_m +$$
$$\nabla \vec{v}_m{}^T)] + \rho_m \vec{g} + \vec{F} + \nabla \cdot (a_1 \rho_1 \vec{v}_{dr,1} \vec{v}_{dr,1} + a_2 \rho_2 \vec{v}_{dr,2} \vec{v}_{dr,2}) \tag{5}$$

Energy Equation for the Mixture

$$\frac{\partial}{\partial t} (a_1 \rho_1 E_1 + a_2 \rho_2 E_2) + \nabla \bullet (a_1 \vec{v}_1 (\rho_1 E_1 + p)$$
$$+ a_2 \vec{v}_2 (\rho_2 E_2 + p)) = \nabla \cdot (k_{eff} \cdot \nabla T) + S_E) \tag{6}$$

Volume Fraction Equation for the Secondary Phase

$$\frac{\partial}{\partial t} (a_p \rho_p) + \nabla \bullet (a_p \vec{v}_m \rho_p) = -\nabla \cdot (a_p \vec{v}_{dr,p} \rho_p) \tag{7}$$

4.2 Computational grids

As the channel of the whole pump is complex and irregularly twisted, unstructured tetrahedral grids are adopted to the channel of inducers and impellers. The GAMBIT software is adopted to draw grids. The computational domain of the high speed pump with equal-pitch inducer consists of 200,097 nodes, 666,699 unit girds, with long-equal-pitch inducer 202,673 nodes, 680,657 unit girds, with progressive pitch inducer 209,658 nodes, 721,189 unit girds, with two-long and two-short inducer 202,673 nodes, 680,657 unit girds. The quality of the grids is satisfied with the solver's demand. The grids are shown in Fig.5.

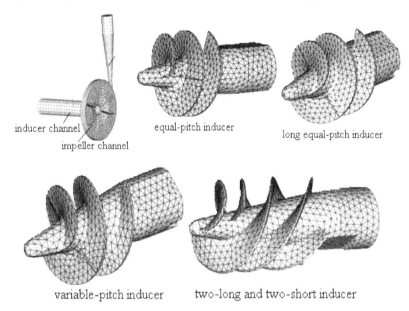

inducer channel
impeller channel
equal-pitch inducer
long equal-pitch inducer

variable-pitch inducer
two-long and two-short inducer

Fig. 5. Computational grids and inducers' grids

4.3 Boundary conditions

1. Inlet. Velocity-inlet is specified on the inlet.
2. Outlet. Static pressure is specified on the outlet. In order to get the distribution of the pressure and the gas-liquid phase volume fraction, the value of the outlet pressure should be the one which will ensure the pump to cavitate. The value is also got by the cavitation performance experiment. It can be seen in the table3.

Inducers	Absolute total pressure/Pa
Equal-pitch inducer	920627.3233
Long-equal-pitch inducer	932401.728
Progressive pitch inducer	932376.728
Two-long and two-short inducer	946138.5334

Table 3. Pressure on the outlet

3. Multi-phase flow. The Mixture model is adopted, and the number of the phases is set as two. The main phase is water-liquid, and the secondary phase is water-vapor. The saturated steam pressure is 3540 Pa.
4. Wall. No slip boundary conditons is specified.
5. Coordinate system. The moving coordinate system is adopted in the channel of the inducer and the impeller, and the rotation speed is set as 6 000 r/min, while the static coordinate system is adopted in the channel of the inlet pipe and the volute.

4.4 Results of numerical simulation

According to above, simulations are done. The velocity distribution and the static pressure distribution are got. For the cavitation mostly depends on the static pressure, the static pressure is chosen to be mainly analyzed. In order to know the pressure distribution mechanism law, the axial profile is chosen to be analyzed, which is shown in Fig.6. The static pressure distribution on the inducer is show in Fig.7.

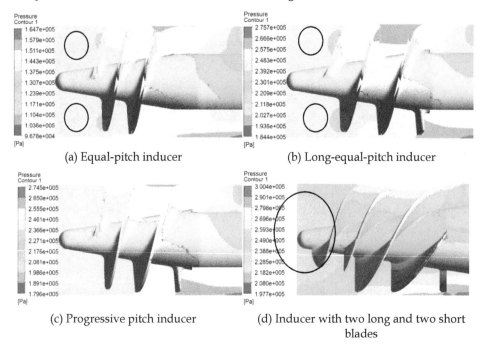

(a) Equal-pitch inducer   (b) Long-equal-pitch inducer

(c) Progressive pitch inducer   (d) Inducer with two long and two short blades

Fig. 6. Static pressure distribution in the axial channel

Fig.6 shows that the static pressure increases gradually from inlet to outlet. The pressure difference between the outlet and th inlet is different. Heads upstream of the impeller can be computed by the pressure difference. Fig.6 shows that near the suction side of the blade low pressure area exists in the equal-pitch inducer, long equal-pitch inducer. The pressure in the inducer's inlet is lower in the two-long and two-short inducer.

In order to know the pressure distribution on the inducers, take the inducers as the research object, which can be seen in Fig.7.

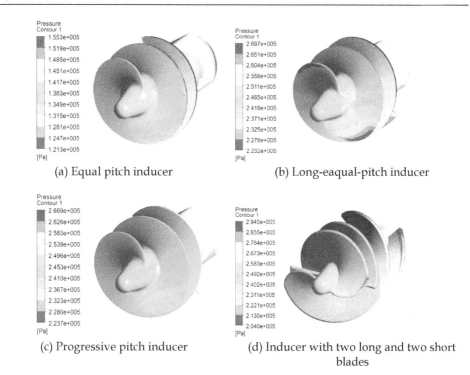

(a) Equal pitch inducer   (b) Long-eaqual-pitch inducer

(c) Progressive pitch inducer   (d) Inducer with two long and two short
                                               blades

Fig. 7. Static pressure distribution on the inducer

Inducers	Absolute static pressure distribution range/Pa
Equal-pitch inducer	121300~155300
Long-equal-pitch inducer	223200~269700
Progressive pitch inducer	223700~266900
Two-long and two-short inducer	204000~294500

Table 4. Static pressure distribution range

Fig. 7 shows that under the design work condition, the static pressure increases gradually from inlet to outlet.The pressure difference between the outlet and the inlet can be got by the simulation. Heads can be computed by the pressure difference, and the result is listed in the table 5.

Inducers	Head of the high-speed centrifugal pump /m
Equal-pitch inducer	97.01
Long-equal-pitch inducer	98.21
Progressive pitch inducer	98.12
Two-long and two-short inducer	98.90

Table 5. Head of the high-speed centrifugal pump

Table 5 shows that the head of the high-speed centrifugal pump is the highest with the two-long and two-short inducer. Second is with long equal-pitch inducer. Third is with progressive pitch inducer. Fourth is with equal-pitch inducer. This is mainly relevant to the helical pitch (L, which can be seen in Table 1). The helical pitch of the two-long and two-short inducer is about 52 mm, long-equal-pitch 37 mm, progressive pitch 33 mm, equal-pitch 20 mm.

$NPSH_r$ can be computed by the equation 8.

$$NPSHr = \frac{v_0^2}{2g} + \frac{\lambda w_0^2}{2g} \tag{8}$$

Where

$$\lambda = 1.2tg\beta_0 + (0.07 + 0.42tg\beta_0)(\frac{s_0}{s_{max}} - 0.615)\frac{-b \pm \sqrt{b^2 - 4ac}}{2a} \tag{9}$$

where,

v_0 – average velocity slightly before the vane inlet.
w_0 – average relative velocity near slightly before the vane inlet.
λ – blade inlet pressure drop coefficien.
β_0 – relative flow angle of the front cover flow lines.
S_0, S_{max} – width of the vane inlet and the max width

According to the simulation results, v_0 and w_0 can be got, and combining with the equation 8 and 9, $NPSHr$ can be computed. The results are show in table 6.

Inducers	$NPSH_r$ of the high-speed centrifugal pump /m
Equal-pitch inducer	0.5910
Long-equal-pitch inducer	0.2624
Progressive pitch inducer	0.3450
Two-long and two-short inducer	0.3691

Table 6. $NPSH_r$ of the high-speed centrifugal pump

Table 6 shows that the centrifugal pump has best cavitation performance when it is with the Long-equal-pitch inducer. Second is with Progressive pitch inducer. Third is with Two-long and two-short inducer. Fourth is with Equal-pitch inducer. This influence order on the cavitation is not same with the influence on the head. The pump with two-long and two-short inducer has highest head, but the cavitation is not the best. The reason is that the inducer is with four vanes, and the extruding coefficient is increased.

5. Cavitation performance experiment

In order to identify the cavitation performance of the pump with four different inducers, the external performance experiments are carried out. The experiment equipment is shown in

Fig. 8. And the test pump is shown in Fig.9. The test inducers are shown in Fig.10. The pump's performance curves under the design point are shown in Fig.11.

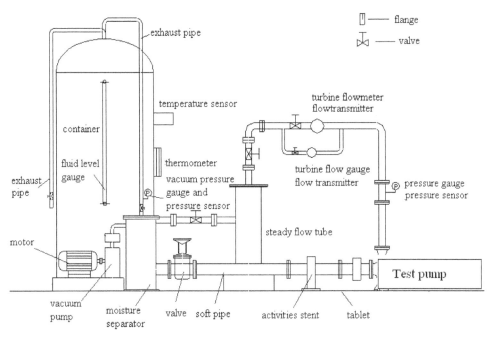

Fig. 8. The experiment equipment

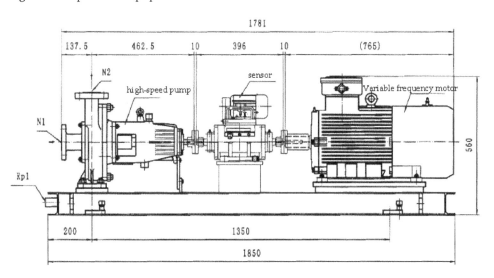

Fig. 9. The test pump

The parameters of the high-speed centrifugal pump are described as previous. The sensor's rated torque is 100N·m. The operation range of the speed of rotation is from 0 to 10000 rpm. The variable frequency motor's rotation is from 0 to 9000r/min，and its max power is 22kw. It is controlled by an inverter.

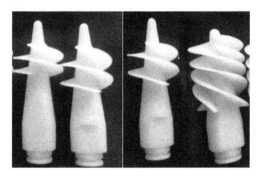

Fig. 10. Test inducers

Test inducers are made by the rapid prototyping. The inducers are respectivly equal-pitch inducer, long equal-pitch inducer, progressive pitch inducer, and the inducer with two long and two short blades.

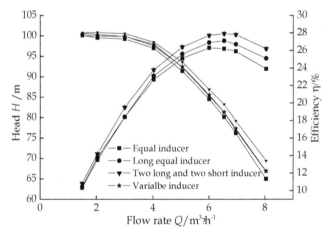

Fig. 11. External performance curves

The H-Q performance curve has no positive slope whether it is with any inducer. On the design work condition, the heads and efficiencies are listed in table 7.

By the contrast of the head in the Table 7 (got by the experiment) and Table 5 (got by the simulation) , it shows that the two values are very close, and has the same law. The pump has highest head when it is with the two-long and two-short inducer, second is with the long-equal-pitch inducer, third is with the progressive pitch inducer, and fourth is with the equal-pitch inducer.

Fig.12 shows head variation with the decrease of the inlet pressure

Inducers	Head /m	Efficiency/%
Equal-pitch inducer	96.97	22.7
Long-equal-pitch inducer	97.96	23. 8
Progressive pitch inducer	97.58	23. 1
Two-long and two-short inducer	98.45	23.8

Table 7. Heads and efficiencies of the high-speed centrifugal pump

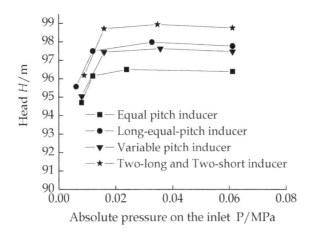

Fig. 12. Head variation with the decrease of the inlet pressure

With the decrease of the inlet pressure, the head of the pump will decline suddenly. From Fig.12, the critical point can be got, and the value is listed in table 8.

Inducers	Absolute pressure on the inlet P/Pa
Equal-pitch inducer	10103.07357
Long-equal-pitch inducer	8141.311713
Progressive pitch inducer	15988.35915
Two-long and two-short inducer	9067.163601

Table 8. Absolute pressure on the inlet

Figure 13 shows that at low flow rate, the cavitation performance of the equal-pitch inducer is not so good, while the long-equal-pitch inducer is good. At high flow rate, the two-long and two-short inducer has best cavitation performance. While the progressive pitch inducer has good cavitation performance whether at the low flow rate or high flow rate. On the design work condition, the NPSHr values are shown in table 9.

Compared with the values got by the simulation in Table 6, it shows that the NPSHr values are very close. The long equal-pitch inducer has best cavitation performance, second is progressive pitch inducer, third is two-long two-short inducer, and last is equal-pitch inducer.

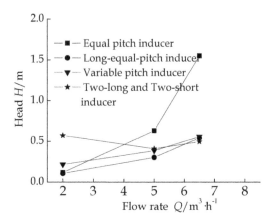

Fig. 13. *NPSHr-Q* curve

Inducers	$NPSH_r$ of the high-speed centrifugal pump / m
Equal-pitch inducer	0.6305
Long equal-pitch inducer	0.3026
Progressive pitch inducer	0.3852
Two-long and two-short inducer	0.4090

Table 9. $NPSH_r$ of the high-speed centrifugal pump

6. Conclusion

The flow of the centrifugal pump with inducers which are respectively with equal-pitch, long-equal-pitch, progressive pitch, two-long and two-short blades are numerically simulated. The corresponding external performance experiment is carried out. From the above, the conclusions can be got as follows:

1. The comparison of the simulation and experiment shows that the trend of every performance curve is similar. For design work conditions, the results obtained from the simulation and experiment are close.
2. The high-speed pump with different inducers has different heads. The head of the high-speed centrifugal pump reaches its highest with two long and two short inducers. The second highest head is achieved with a long equal-pitch inducer. The third highest is realized with the variable pitch inducer, and the fourth is achieved with an equal pitch inducer.
3. Adding an inducer can improve pump cavitation performance. The long equal pitch inducer exhibits the best cavitation performance; the second is the progressive pitch inducer; the third is the device with two long and two short inducers, and the last is the equal pitch inducer.
4. The pump with an inducer's head is mainly relevant to the helical pitch L. So when design inducer, the helical pitch L should be longer appropriately.
5. The research can supply significant guide for inducer's design.

7. Appendix

Notation

a_1, a_2	liquid-phase/gas-phase volume fraction
F	volume force (N)
H	head (m)
$\dot{m}$	cavitation effect of mass transfer
n	rotation speed (r min^{-1})
Q	flux (m^3 h^{-1})
v	mass mean velocity (m s^{-1})
v_1, v_2	liquid-phase/gas-phase velocity (m s^{-1})
ρ	mixture density (kg m^{-3})
ρ_1, ρ_2	liquid-/gas-phase density (kg m^{-3})
μ_1, μ_2	liquid-/gas-phase dynamic viscosity (pa ·s)

8. Acknowledgment

This project is supported by National Science and Technology Support (50879080, 50976105), ZheJiang Science and Technology Support (Y1100013)

9. References

Ait-Bouziad, Y., Farhat, M., Guennoun, F., Kueny, J. L., Avellan, F., and Miyagawa, K. Physical modelling and simulation of leading edge cavitation: application to an industrial inducer. *Fifth International Symposium on Cavitation*, 2003, Osaka, Japan.

Ait-Bouziad, Y., Farhat, M., Kueny, J. L., Avellan, F., and Miyagawa, K. Experimental and numerical cavitation flow analysis of an industrial inducer. *22th IARH Symposium on Hydraulic Machinery and Systems*, 2004, Stockholm,Sweden.

Benoît Pouffary, Regiane Fortes Patella, Jean-Luc Reboud, Pierre-Alain Lambert. Numerical simulation of 3d cavitating flows: analysis of cavitation head drop in turbomachinery. *ASME J. Fluids Eng.*, Vol.130, (June 2008), pp. 061301.1-10. ISSN 0021-9223

Benoît Pouffary, Regiane Fortes Patella, Jean-Luc Reboud, and Pierre-Alain Lambert. Numerical analysis of cavitation instabilities in inducer blade cascade. *Journal of Fluids Engineering*, Vol. 130 , (April 2008), 041302-1-8. ISSN: 0021-9223

Cui Bao-lin, Chen Ying, and Zhu Zuchao. Numerical simulation and theoretical analysis of high-speed centrifugal pump with inducer. Hangzhou: Zhejiang university, 2006.

Coutier-Delgosha, O., Courtot, Y., Joussellin, F., and Reboud, J. L.. Numerical simulation of the unsteady cavitation behavior of an inducer blade cascade. *AIAA J.*, Vol 42, No. 3, (2004), pp:560–56.

Ding Xi-ning, LIANG Wu-ke. Numerical simulation of two-phases cavitation flow in equal-pitch inducer. *Journal of Water Resources and Water Engineering*. Vol.26, No.5, (December 2009),pp.170-172. ISSN 1672-643X

Fortes Patella, R., Coutier-Delgosha, O., Perrin, J., and Reboud, J. L. A numerical model to predict unsteady cavitating flow behaviour in inducer blade cascades. *ASME J. Fluids Eng.*, Vol.129, No 1, (2007), pp: 128–135. ISSN 0021-9223

Guo Xiaomei, ZHU Zuchao, and CUI Baoling, Analysis of cavitation and flow computation of inducer. *Journal of Mechanical Engineering*. Vol.46, No.4, (April 2010), pp.122-128, ISSN 0577-6686

Hosangadi, A., and Ahuja, V. Simulations of cavitating flows using hybrid unstructured meshes. *ASME J. Fluids Eng.*, 2001, 123, pp. 331–340. ISSN: 0021-9223

Hosangadi, A., Ahuja, V., and Ungewitter, R. J. Numerical study of a flat plate inducer: comparison of performance in liquid hydrogen and water. *Sixth International Symposium on Cavitation*, CAV2006, Wageningen, September, 2006. The Netherlands

Kong Fanyu, Zhang Hongli, Zhang Xufeng, and Wang Zhiqiang. Design on variable-pitch inducer based on numerical simulation for cavitation flow. *Journal of Drainage and Irrigation Machinery Engineering*, Vol.28, No.1, (January 2010), pp:12-17, ISSN 1674-8530

Kunz, R. F., Boger, D. A., Stinebring, D. R., Chyczewski, T. S., Lindau, J. W., and Gibeling, H. J. A preconditioned navier–stokes method for two-phase flows with application to cavitation. *Comput. Fluids*, 29(8), 2000, pp:849–875. ISSN 0045-7930

Langthjem M A and olhoff N. A numerical study of flow-induced noise in a two-dimensional centrifugal pump. Part L Hydrodynamics. *Journal of Fluids and Structures*, 2004（19）pp:349-368, ISSN 0889-9746

Li W G. Effect of volute tongue on unsteady flow in a centrifugal pump. *International Journal of Turbo & Jet Engines*, 2004（21）pp:223-231, ISSN 0334-0082

Li Yaojun, and Wang Fujun. Numerical investigation of performance of an axial-flow pump with inducer. *Journal of Hydrodynamics*. 2007,9(6): 705-711. ISSN：1001-6058

Medvitz, R. B., Kunz, R. F., Boger, D. A., Lindau, J. W., Yocum, A. M., and Pauley, L. L. Performance analysis of cavitating flow in centrifugal pumps using multiphase CFD. *ASME-FEDSM*, 2001,01, New Orleans

Mejri, I., Bakir, F., Rey, R., and Belamri, T. Comparison of computational results obtained from a homogeneus cavitation model with experimental investigations of three inducers. *ASME J. Fluids Eng.*, 2006, 128, pp.1308–1323. ISSN: 0021-9223

OkitaK, UgajinH, and MatsumotoY. Numerical analysis of the influence of the tip clearance flows on the unsteady cavitating flows in a three-dimensional inducer. *Journal of Hydrodynamics*, Vol 21, No.1, (2009）pp:34-40, ISSN 1001-6058

Tang Fei, Li Jiawen, Chen Hui, LI Xiangyang, and Xuan Tong. Study on cavitation performance of inducer with annulus inlet casing. Journal of Mechanical Engineering, Vol 47, No.4, (February 2011), pp:171-176, ISSN 0577-6686

Wang Jian-ying, and Wang Pei-dong. Application of screw inducer used in high-performance centrifugal pump. *Gas Turbine Experiment and Research*. Vol.19, No.2, (May 2006), pp.43-46, ISSN 1672-2620

Yuan Dan-qing, Liu Ji-chun, Cong Xiao-qing, and Wang Guan-jun. Numerical calculation of cavitation for inner flow field of variable-pitch inducer. *Drainage and Irrigation Machinery*, Vol.26, No.5, (August 2008), pp:42-45. ISSN 1674-8530

Impeller Design Using CAD Techniques and Conformal Mapping Method

Milos Teodor

"Politehnica" University of Timisoara,
Romania

1. Introduction

Computerized pump design has become a standard practice in industry, and it is widely used for both new designs as well as for old pumps retrofit. Such a complex design code has been developed over the past decade by the author. However, any design method has to accept a set of hypotheses that neglect in the first design iteration the three-dimensional effects induced by the blade loading, as well as the viscous effects. As a result, an improved design can be achieved only by performing a full 3D flow analysis in the pump impeller, followed by a suitable correction of the blade geometry and/or the meridian geometry. This chapter presents a fully automated procedure for generating the inter-blade channel 3D geometry for centrifugal pump impeller, starting with the geometrical data provided by the quasi-3D code. This is why we developed an original procedure that successfully addresses all geometrical particularities of a centrifugal pump.

2. Domain generation of axial-symmetric flow in impeller area

Fluid movement in the impeller area is axially symmetric. The reference system adequate to this kind of flow is cylindrical (r,θ,z). Because of geometric axial symmetry of the domain area we have axial-flow symmetry and cinematic, so the study of spatial movement can be reduced to the study of plane movements, or in a meridian plane, plane containing the rotation axis of symmetry, Oz. The resolution must be analytical in order to continue to analytically generate the mesh network to simulate the flow by Finite Element Method (FEM). The inlet data are the main dimensions of the preliminary study. First axial and radial extensions should be set. Axial extension, fig. 1, shall be determined based on previous studies (Gyulai, 1988) relation:

$$z_{\max} = 1,1 \cdot D_0 \tag{1}$$

Radial expansion is taken by 25% over D_{2ex} diameter (D_2-extended). At the mixed-flow-impellers, impeller diameter output, D_2, is in the area of transition from axial to radial movement. Therefore the domain must be extended to at least $2D_0$. So if $D_2 < 2D_0$ then $D_{2ex}=2D_0$, and if $D_2 \geq 2D_0$, then $D_{2ex}=D_2$. From D_{2ex} we calculate D_{2max} with the relationship:

$$D_{2\max} = 1,25 \cdot D_{2ex} \tag{2}$$

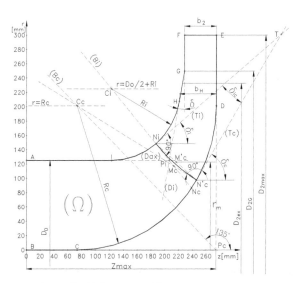

Fig. 1. Geometric construction of half-domain of flow

Z_{max} and D_{2max} define dimensions of the gauge domain. Next axial area with radial area is connected. Because of technological reasons the shroud connection is made with a straight segment GH and an arc of radius R_i and at the hub only an arc of radius R_c. Input section AB is determined by the input diameter D_0 and output section EF of diameter D_{2max} and width b_2. First the shroud connection is done by and depending of its position is determined the hub radius R_c and its centre position. The starting point is G at $D_{2G}=1,05D_{2ex}$. D_{2G} is placed by 5% above D_{2ex} because if $D_{2ex} = D_2$ then in the vicinity of G appear some inflections in meridian speeds variation which will distort the speed triangles of output. From G goes straight (Di) tilted from the vertical at angle $\delta = 2° \ldots 8°$. Angle δ is chosen in an initial approximation for n_q-low around 2°, and for n_q-high values close to 8°. Many solutions are possible within the range specified; it is chosen the one taking into account other considerations, then calculating the required width at D_{2ex} diameter. It is usually opted for a rounded value, and after obtaining velocity variation along the streamlines the calculation can be redone with other values of the tilt angle δ of conical area.

From A doing a parallel to Oz right results straight (Dax), which intersected with (Di) determines the point P_i. The centre of connecting arc HI will be on bisector of angle. R_i is chosen depending on the type of impeller, for slow impellers, n_q-low, R_i is small, and for fast impellers, n_q-high, is high because the impeller's blades will be mainly in the area of curvature (crossing).

Exact coordinates of point of contact H result from the analytical condition of tangent for the straight taken from point G to the arc connecting the shroud, for all using polar equation and tangent equation to a circle of analytical geometry.

Shroud area is completely defined, the following step will be to determine the hub radius of the arc connecting to the passage so that the section area of curvature equals to the input section. Centre of arc connecting the hub will be on bisector (BC) of right angle $C\hat{P}_cD$.

Finding the optimal connection radius is done through repeated testing, starting with low values until the condition of equal sections is verified with an error less than 1%.

The passage sectional area is calculated and verified in the end by the relationship:

$$A_m = 2\pi r_m b_m = \frac{\pi D_0^2}{4} \tag{3}$$

This way the domain is completely defined analytically and with the exception of point G, all connections are continuous and crossing sections from A to G are relatively constant.

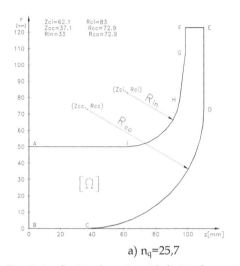 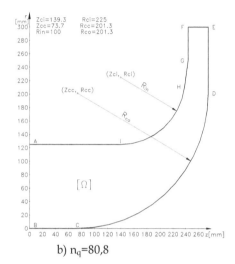

a) $n_q=25,7$ b) $n_q=80,8$

Fig. 2. Analyzing domain with finite elements

3. Domain discretization

Integrating flow function ψ and potential velocity function φ is done by FEM using quadrilateral izo-parametric linear finite elements. To maintain good accuracy in the application of FEM it is necessary that the sides of quadrilaterals to be relatively equal in length, that some sides are not disproportionately small compared to others.

Since the average length and width ratio of the domain is about 10, 100 intervals were taken between input and output and 10 intervals in the cross section. Borders shroud and hub were divided into 100 equal intervals resulting in 101 points (Fig. 3). Homologous points on the shroud and hub were joined together by line segments. Each straight segment in turn was divided into 10 intervals, resulting in 11 points on each segment. Number of nodes of the mesh was 101 * 11 = 1111, every 4 neighbouring nodes create an irregular quadrilateral that will be the base of integration for stream function Ψ in partial differential equation.

This method is recommended for the case when the mesh is an exercise in applying FEM. Flow professional software automatically solves this problem after the domain geometry is defined.

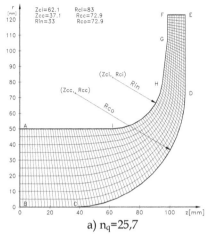
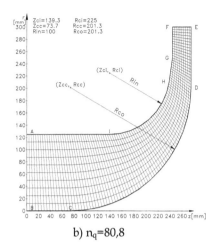

a) $n_q=25,7$ b) $n_q=80,8$

Fig. 3. Network of the finite element mesh

4. Hydrodynamic field of the impeller

Partial differential equation of the stream function ψ is a Helmholtz equation of elliptic cylindrical coordinates:

$$\frac{\partial^2 \psi}{\partial z^2} + \frac{\partial^2 \psi}{\partial r^2} - \frac{1}{r}\frac{\partial \psi}{\partial r} = 0 \tag{4}$$

Flow differential function is:

$$d\psi = \frac{\partial \psi}{\partial r}dr + \frac{\partial \psi}{\partial z}dz = rv_z dr - rv_r dz \tag{5}$$

where velocity components are:

$$v_z = \frac{1}{r}\frac{\partial \psi}{\partial r} \quad ; \quad v_r = -\frac{1}{r}\frac{\partial \psi}{\partial z} \tag{6}$$

Integrating this equation is done by FEM using the real values of coordinates of the points of the mesh network, expressed in mm. Dimensionless treatment has been dropped because of the method in which concrete cases are solved for the later stages using real values of velocities. Solving the global FEM system is better than unknowns, values of function ψ to be of the same order of magnitude with coefficients given by the coordinates of points in order to increase accuracy of calculation. Therefore ψ admits border values between 0 and 100.

5. The boundary conditions on domain border

Borders $AIHGF$ and $BCDE$ are streamlines and ψ=const along them. We recognize on AF boundary ψ=100 and on BE, ψ=0. Thus on input boundary we have a uniform flow with constant velocity on the entire section:

$$v_r = 0 \quad ; \quad v_z = \frac{4Q}{\pi D_0^2} \tag{7}$$

Replacing (7) in (5) and integrating results:

$$\psi = \int \frac{4Q}{\pi D_0^2} r dr = \frac{4Q}{2\pi D_0^2} r^2 + C_i \tag{8}$$

Admitting that on the hub border we have $\psi = 0$, including the point B, will result integration constant C_i:

$$(r = 0) \quad \Rightarrow \quad (\psi = 0) \quad \Rightarrow \quad (C_i = 0) \tag{9}$$

On the shroud border stream function will be constant and equal to that of point A where $r = D_0/2$, and $\psi = Q/2\pi = 100$ (quasi-unitary flow).

$$v_r = \frac{Q}{\pi D_{2\max} b_2} \quad ; \quad v_z = 0 \tag{10}$$

Replacing (4.10) in (4.5) and integrating results:

$$\psi = -\int r v_r dz = -\int r \frac{Q}{\pi D_{2\max} b_2} dz \Big|_{r=\frac{D_{2\max}}{2}} = -\frac{zQ}{2\pi b_2} + C_e \tag{11}$$

Admitting the same conditions on the solid boundaries result:

$$(z = z_E) \quad \Rightarrow \quad (\psi = 0) \quad \Rightarrow \quad \left(C_e = \frac{Q z_{\max}}{2\pi b_2}\right) \tag{12}$$

As a result, the law of variation ψ on the border EF is given by:

$$\psi = -\frac{Qz}{2\pi b_2} + \frac{Q z_{\max}}{2\pi b_2} = \frac{Q}{2\pi} \frac{1}{b_2} (z_{\max} - z) \tag{13}$$

Integration with FEM of the Helmholtz equation (4) will be in an area where function values are imposed on border which means that we have to solve a problem of Dirichlet type.

6. Calculus of stream function Ψ by FEM

The above have created all necessary conditions for the Helmholtz equation (4) integration by FEM. Function ψ can be globally approximated on Ω by:

$$\psi = a_\alpha \psi_\alpha \quad ; \quad \alpha = \overline{1, G} \tag{14}$$

where G is the number of nodes on Ω. Applying Galerkin's method (Anton at al. 1988) follows:

$$\int_\Omega \left(\frac{\partial^2 \psi}{\partial z^2} + \frac{\partial^2 \psi}{\partial r^2} - \frac{1}{r} \frac{\partial \psi}{\partial r} \right) a_\alpha d\Omega = 0 \tag{15}$$

Using the notation (Anton at al. 1988) resolution is reduced to solving the global linear system of equations:

$$D_{\alpha\beta}\Psi_\beta = F_\alpha \quad \text{where} \quad \alpha,\beta = \overline{1,G} \tag{16}$$

The method of solving the system can be Gauss-Seidel iterative method, resulting in final values of the stream function ψ mesh nodes. From a mathematical point of view the lines are defined by the geometrical locus of points where the stream function has the same values. If between the solid borders the stream function ψ takes values between 0 and 100, then streamlines having ψ = 10, 20,..., 90 are looked for because ψ = 0 and ψ = 100 are the solid borders. Identification of the points for stream lines is made through interpolation with cubic SPLINE function to have more precision.

Applying the same methodology (Anton at al. 1988) as for the stream function ψ can integrate the equation for the velocity potential function, φ, in the end getting equipotential lines that overlap the stream lines as shown in Fig. 4.

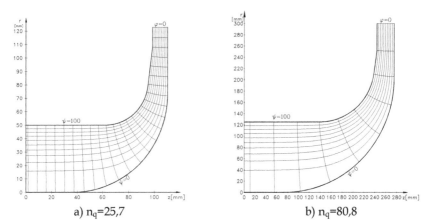

a) n_q=25,7   b) n_q=80,8

Fig. 4. Stream and equipotential lines of the hydrodynamic spectrum

7. Determination of the velocities and pressures fields

Taking into account relations (6) and the notation (Anton at al. 1988) meridian velocity components in the centre of gravity of each finite element are calculated with relations:

$$\begin{cases} v_z^e = \dfrac{4}{\alpha_0 a_2} A_{N2}\psi_N^e \\[2mm] v_r^e = \dfrac{4}{\alpha_0 a_2} A_{N1}\psi_N^e \end{cases} \tag{17}$$

Meridians speed module will be calculated with the relationship:

$$v_m^e = \sqrt{\left(v_z^e\right)^2 + \left(v_r^e\right)^2} \tag{18}$$

Speed on borders is calculated by extrapolation. When calculating the pressure a Bernoulli equation is applied along a stream lines, between a point on the inlet border and a current point on the domain, points belonging to the same stream line. If on the boundary AB velocity is constant and equal with v_0 and pressure p is p_0 we have the Bernoulli equation:

$$p = p_0 + \frac{\rho}{2}\left(v_0^2 - v_m^2\right) \tag{19}$$

Dividing the current speeds and pressures with p_0 and v_0 so that we form dimensionless calculation:

$$\overline{v}_m = \frac{v_m}{v_0} \tag{20}$$

$$\overline{p} = C_p = \frac{p - p_0}{\rho \dfrac{v_0^2}{2}} = 1 - \left(\overline{v}_m\right)^2 \tag{21}$$

Figures 5 and 6 present the speeds variation in meridian plane along the stream lines, respectively the pressure coefficient C_p for the two types of impellers.

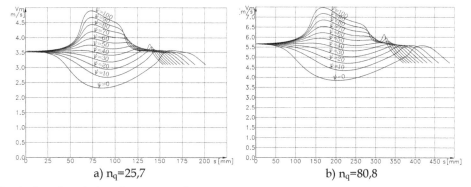

Fig. 5. Speed variation along stream lines

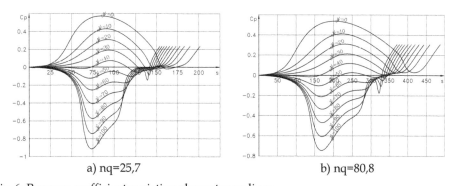

Fig. 6. Pressure coefficient variation along stream lines

8. Choosing the blade area in the domain of hydrodynamic field

Classic calculus relations and design method are combined with facilities of computer use in this phase. Meridian hydrodynamic field data in an optimized domain depending of impeller type offers the perspective of results close to reality. Sections of the calculation are equal to the number of stream lines (11 lines of flow) which means doubling or even tripling them to the case of using the graphic-analytical method of tracing the hydrodynamic field.

Choice of blade area in as many variations or options initially imposed is an additional possibility to optimize the blades.

From the preliminary study it is known where D_2 will be located on the output edge. In the meridian plan will be a straight with inclined angle γ_2 (Fig. 7). If $n_q \leq 35$ is recommended that $\gamma_2 = 0$, and for $n_q > 35$ is acceptable so that it can tilt somewhat orthogonal to the middle flow line. Approximate statistical relationship can be used, (Gyulai, 1988):

$$\gamma_2 = 44,037 \lg n_q - 68,213 \tag{22}$$

The output edge pivots around point P_{me} at radius $r_2 = D_2/2$ on the medium stream line. The final decision on the angle γ_2 is taken after reasons that follow.

At the inlet there is an infinitely of possible solutions. Angle is also given guidance by (Gyulai, 1988):

$$\gamma_1 = 55,19 \lg n_q - 53,726 \tag{23}$$

With the marked output edge marked image from fig. 7 is obtained; in which on the shroud edge are given the points of the contour mesh. Choose the position of P_{mi} (the pivot edge inlet) so that there is enough space to carry blades because otherwise it will result in a disproportionate number of blades. Because the inlet edge is situated in the curvature of the hydrodynamic field is necessary correlate its position with extreme speed values near the shroud because as meridians are maximum and speed of transport u_1 is also maximum and at the hub the situation is reversed in the sense of extreme minimum.

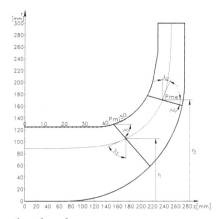

Fig. 7. Marking the inlet and outlet edges

These two extreme speed values involve the angle β'_1 whose values will be slightly different between the shroud and hub. Only the blades zone is retained from the stream lines domain and speeds and pressures. From this moment the rest of the hydrodynamic field doesn't interest us in terms of usefulness for the blade zone, only the edges between input and output. At this stage any option may be followed by a rerun, if the image of velocity variation in the blade (Fig. 8) does not meet the target. Next we are interested only in the hydrodynamic field of the impeller area where we want to avoid areas at risk for increased sensitivity to cavitation or emphasis of the degree of blockage. Therefore after a first option to input the input and output edges, the variation of speeds along the stream lines is displayed again with markings for the blade area (bold in Fig. 8). It is generally better that the input edge is in the extremes of speed curves or after in the flow direction. The justification is given by two aspects:

1. after the stream inlet in the blades zone is good for speeds to have tended to decrease rather than increase, which would increase the risk of cavitation.
2. height of speed triangles from input (Fig. 10) is affected by the values of these speeds and by these the construction blade angle of the inlet, i.e. high levels of transport speed, u_1, (SL shroud) correspond to proportional values for the v_m and vice versa to the hub, so the angle does not fall below $10°$.

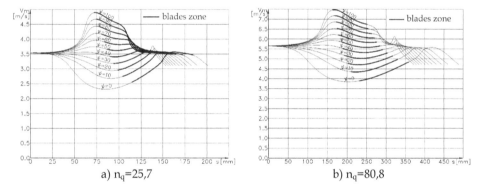

a) $n_q=25,7$ b) $n_q=80,8$

Fig. 8. Marking changes in velocity in the blade

If the angle falls below $10°$ there will be problems with the degree of blockage in an inlet in that $\rho_1 < 0.6$, which is again unacceptable because of increased susceptibility to cavitation.

After the final decision on the position of input and output edges, the meridian velocity variation along of input and output edges is presented. These speeds will be used to calculate kinematic and angular elements of inlet and outlet, which offer height of speed triangle.

9. Blade inlet design

Initial sizing data for inlet are meridian speeds on the inlet edge at the intersection points of stream lines with inlet edge (radiuses resulted), blade thickness (minimum 4 mm), transport speeds (tangential) at the point of calculation: $u_1 = r_1\omega$. The number of blades is determined by statistical relationship (Pfleiderer, 1961):

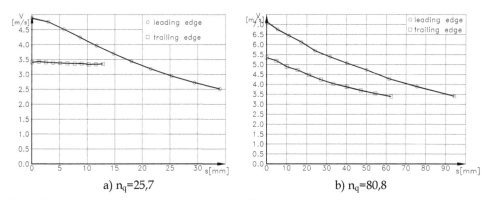

a) n_q=25,7   b) n_q=80,8

Fig. 9. Velocity variation along the meridians of input and output edges

$$z_1 = 4\pi \sin^2 \beta_m \cos \lambda_m \frac{r_2 + r_1}{r_2 - r_1} \tag{24}$$

where $\beta_m = \dfrac{\beta_1 + \beta_2}{2}$, $\lambda_m = \dfrac{\lambda_1 + \lambda_2}{2}$ r_1, r_2, λ_1, λ_2 (fig. 7.) are data related to medium stream line. Angle β_2 is calculated in a first approximation by the statistical relationship, (Gyulai, 1988):

$$\beta_2 = \frac{106}{n_q^{0,235}} \tag{25}$$

and β_1 is the angle of inlet without blockage given by the thickness of the blade

$$\beta_1 = arctg \frac{v_{m1}}{u_1} \tag{26}$$

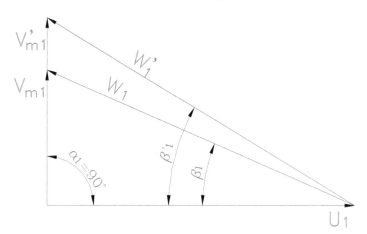

Fig. 10. Velocity triangles at the impeller inlet

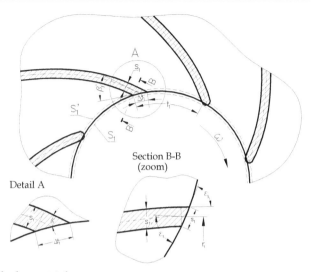

Fig. 11. Blades blockage at inlet

The essential condition for inlet is the alignment of the calculated blade direction to the stream direction given by the inlet speeds triangle. Blades by their inclined position in the impeller channel and their thickness causes a decrease of the normal passing section through the entire blades zone. The measure of this reduction is given by the degree of blockage which in geometric terms is the ratio of the blocked section and unobstructed section. To this effect the degree of blockage, ρ_1, given the thickness of the blades s_1, must be taken into account:

$$\rho_1 = \frac{S_1'}{S_1} = \frac{b_1(t_1 - \Delta t_1)}{b_1 t_1} = 1 - \frac{\Delta t_1}{t_1} = 1 - \frac{\Delta t_1}{\dfrac{2\pi r_1}{z_1}} \tag{27}$$

In figure 11 is observed that Δt_1 is blade thickness on the inlet section and on the stream surface. Reference thickness, s_1, based on the hydrodynamic flow through the impeller is thickness measured in a direction normal to the blade contour from the stream surface. If β_1' is blade construction angle of inlet then from the inlet speed triangle can be written as:

$$\Delta t_1 = \frac{s_1}{\sin \beta_1'} \tag{28}$$

Substituting in (27) one gets the calculating relation of the degree of blockage of the fluid depending on the thickness measured on the stream surface:

$$\rho_1 = 1 - \frac{z_1 s_1}{2\pi r_1 \sin \beta_1'} \tag{29}$$

Real thickness, s_{1r}, of the blade is the thickness measured perpendicular to the blade surface and it counts in the calculation of resistance and the formation of the casting model. It is highlighted under section B - B in Fig. 11 and is calculated based on the tilting of the blade surface to the plane tangent to the flow surface at the point of view, tilt angle estimated by ε_1, so that:

$$s_{1r} = s_1 \cdot \sin \varepsilon_1 \qquad (30)$$

The degree of blockage depending on real thickness is calculated by the relationship:

$$\rho_1 = 1 - \frac{z_1 s_{1r}}{2\pi r_1 \cdot \sin \beta_1' \cdot \sin \varepsilon_1} \qquad (31)$$

Relation (31) resulted in geometric terms. Using a result of the continuity relationship (flow volume) will result the link of the blockage degree with the cinematic elements of inlet, (speeds). Having no ante-stator, normal inlet is considered, $\alpha_1 = 90°$, (Fig. 10). Meridian speed with blockage is calculated under the formula:

$$v'_{m1} = \frac{v_{m1}}{\rho_1} \qquad (32)$$

then the resulting angle of blades construction at inlet:

$$\beta_1' = arctg\frac{v'_{m1}}{u_1} = arctg\frac{tg\beta_1}{\rho_1} = arctg\frac{tg\beta_1}{1 - \dfrac{z_1 s_{1r}}{2\pi r_1 \cdot \sin \beta_1' \cdot \sin \varepsilon_1}} \qquad (33)$$

If in relation (33) v'_{m1} is replaced with (32) or ρ_1 with (31) we get a default formula for β_1'. Its solution is obtained through an iterative calculation initiated by an approximated value of ρ_1 ($\rho_1 = 0.8$). The resulting array is fast converging to the sought value.

For centrifugal pumps with low or middle n_q the thickness of the blades is usually constant between inlet and outlet. Only at the inlet rounding is practiced by a circle or ellipse arc so as to not have one of the cases marked NO in Fig. 12. These conditions favour flow separation at the inlet and thus favour cavitation occurrence.

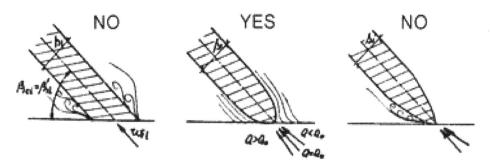

Fig. 12. Blade inlet shape (Gyulai, 1988)

10. Blade outlet design

Initial blade outlet data are meridian speeds in edge points of the outlet edge, radiuses of these points, blade thickness (minimum 4 mm), transport speed in such points $u_2 = \omega r_2$. At outlet occurs the deflection effect of the flow from the direction of blade. Quantifying it is the coefficient of Pfleiderer, p (Pfleiderer, 1961):

$$p = k_p \left(1 + \frac{\beta_2^{[\circ]}}{60} \right) \frac{r_2^2}{z_2 S} \tag{34}$$

where $S = \int_{x_1}^{x_2} r dx$ is static moment of the arc-line between inlet and outlet of the streamline. Streamline is given by points and for the numerical calculation of the integral the summation is used:

$$S = \int_{x_1}^{x_2} r dx \cong \sum_{i=1}^{I_{max}} \left(\frac{r_i + r_{i-1}}{2} (x_i - x_{i-1}) \right) \tag{35}$$

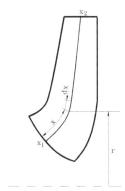

Fig. 13. Geometric schema for static momentum for one streamline

In relation (35) summation is done on small arcs that approximate the centre of gravity of the bowstring. For coefficient k_p values recommended by what is following the impeller in three different cases:

- $k_p = 1 \dots 0.85$ for impeller followed by a space without blades
- $k_p = 0.85 \dots 0.65$ for impeller followed by collector
- $k_p = 0.65 \dots 0.60$ for impeller followed by stator blades

Assuming infinite number of blades: $z_2 \to \infty$, flow deviation tends to zero. Thus the outlet will consider two triangles of speeds (Fig. 14), Δ_2 corresponding to the geometrical construction of blade, and for the real fluid motion with deflection, triangle Δ_3.

Hydraulic moments corresponding to the two situations are in the ratio:

$$\frac{M_{h\infty}}{M_h} = \frac{\rho Q_t (r_2 v_{u2} - r_1 v_{u1})}{\rho Q_t (r_2 v_{u3} - r_1 v_{u1})} = 1 + p \tag{36}$$

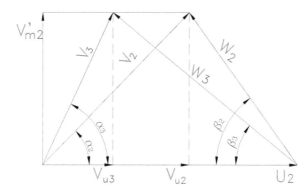

Fig. 14. Velocity triangles at the impeller outlet

Outlet energy transfer depends on three parameters with three degrees of freedom thus:

1. outlet impeller diameter, D_2
2. construction angle of blade, β_2
3. number of blades out z_2 (implied network density)

Simultaneous optimization of all three options is very difficult. Therefore, in the first phase, for the diameter D_2 the initially determined value by the head coefficient, ψ_H, is accepted and for the other two several possible variants are combined. Number of blades is stored as options calculated initially at inlet. If that number doubles or triples this is taken into account, and outlet options will be double or triple than inlet. For β_2 first is tested the rounded expected value, given by (25), then depending on results are also allocated other possible variants near baseline until achieving the conditions imposed by check, (Gyulai, 1988):

a. achieve of theoretical pumping head, H_t
b. the condition of optimum solidity ($l/t_m \cong 2$)

Most important relations used in the calculations (except those relating to the results of solving the kinematic and angular elements through solving triangles of speeds) are:

-   Theoretical pumping head H_t:

$$H_t^{'} = \frac{1}{g} u_2 v_{u3} \geq H_t \quad \text{(from preliminary calculation)} \tag{37}$$

-   Theoretical pumping head assuming infinite number of blades $H_{t\infty}$:

$$H_{t\,\infty} = \frac{1}{g} u_2 v_{u2} \tag{38}$$

-   The relationship between the two pumping heads:

$$\frac{H_{t\,\infty}}{H_t^{'}} = \frac{v_{u2}}{v_{u3}} = 1 + p \tag{39}$$

Outlet calculation is a calculation to check the "n" variants out of which the optimal variant is chosen as the best option that is closest to the initial conditions, i.e. achieving theoretical pumping head and the condition of optimum solidity ($\cong 2$).

11. Calculation of the blade surface in the space frame between inlet and outlet using CAD techniques

The route of inlet to outlet is resolved by interpolating one of the significant kinematic quantities for load distribution along the inter-blades channel of impeller. Interpolation is done along the streamline (flow 3D surface) controlled by curvilinear coordinate "x" and not by the current radius "r" because it better quantifies the load (load distribution) in the radial-axial zone. Hydraulic momentum is the size that reflects blade loading between inlet and outlet. For a current point of curvilinear coordinate x it says:

$$M_{hx} = \rho Q_t \left(r_x v_{ux} - r_1 v_{u_1} \right) \tag{40}$$

From (40) specifying $(r v_u)_x$ follows:

$$\left(r v_u \right)_x = r_1 v_{u1} + \frac{M_{hx}}{\rho Q_t} = f(x) \tag{41}$$

Size directly related to hydraulic momentum is $(r v_u)_x$, so if the variation of product $r v_u$ along impeller channel is controlled, then results the variation of hydraulic momentum versus radius, partial pumping head distribution, distribution of pressure differences on the faces of the blade. $r v_u$ product variation implies variation of the construction angle β necessary under the assumption that the relative velocity is tangential to the middle surface of the blade. Height of speeds triangles is given by meridian speed, v_m', which is v_m corrected with degree of blockage, $\rho_{1\text{-}2}$, resulted from thickness of the blades. If we consider a current point on streamline denoted by "i", then current angle β_i is resulting from the relationship:

$$\beta_i = arctg \left(\frac{v_{mi}}{\rho_i} \frac{1}{u_i - v_{ui}} \right) \tag{42}$$

It is noted that in (42) appear factors r_i ($u_i = r_i \omega$) and v_{ui}, so no matter which of the two variants are interpolated ($r v_u$ or β), we get the same kind of information. Most often angle β, is chosen, being directly related to the blade channel orientation in impeller.

β angle variation between inlet and outlet must be chosen so that there is a relatively uniform blade loading and the variation is strictly increasing throughout the area. For a better flow engage at the inlet and outlet, as stated above, it is recommended that at neighbourhood extreme points, the blade loading tend to zero. Analyzing several cases of impellers that condition is satisfied if the curve of β variation have derived zero at inlet and outlet. Computer solving is possible only through an analytical generation.

11.1 β angle interpolation with two connected parabola arcs

Connecting the two arcs of parabola is the point x_3 as common tangent (Fig.15). Functions that define two parabola arcs with vertical focal axis noted by f_1 and f_2 with general equations, (Milos, 2009):

$$\begin{cases} f_1(x) = a_1 x^2 + b_1 x + c_1 \\ f_2(x) = a_2 x^2 + b_2 x + c_2 \end{cases} \tag{43}$$

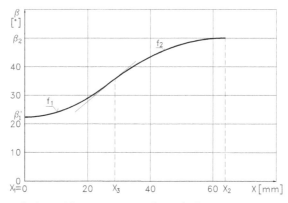

Fig. 15. β angle interpolation with two connected parabola arcs

It is noted that in order to concretely define is necessary to know the coefficients a_1, b_1, c_1, a_2, b_2, c_2. So we need six equations with six unknowns derived from equations (43). Putting analytical positioning and connection conditions we have under notation of Fig. 15:

i.   the two parabola arcs are connected by the same tangent in x_3
ii. the first arc of parabola is tangential to the horizontal (has derivative zero in x_1)
iii. the second arc of parabola is tangential to the horizontal (has derivative zero in x_2)
iv. the two parabola arcs have the same value in x_3
v.   the first arc of parabola passing through the point coordinates (x_1, β_1')
vi. the second arc of parabola passing through the point coordinates (x_2, β_2')

Translating the six analytical conditions will result in a system of six equations with six unknowns (44) is solved exactly by Gauss elimination algorithm.

$$\begin{cases} f_1'(x_3) = f_2'(x_3) & \text{(I)} \\ f_1'(x_1) = 0 & \text{(II)} \\ f_2'(x_2) = 0 & \text{(III)} \\ f_1(x_3) = f_2(x_3) & \text{(IV)} \\ f_1(x_1) = \beta_1 & \text{(V)} \\ f_2(x_2) = \beta_2 & \text{(VI)} \end{cases} \tag{44}$$

Point position x_3 in interval (x_1, x_2) changes step by step until the optimal shape of a blade skeleton variation will result. It is noted that blade load has an infinite range of solutions for each streamline (in 3D surface flow). If each streamline is treated separately charging flow is unlikely to result in a smooth surface (non-sinuous). Blade loading for each streamline must be linked with loads on neighbouring lines. Viewing blade loading by the β angle is necessary but not sufficient. Calculation should be continued until the skeleton surface projection of the blade in a plane perpendicular to the axis of rotation is obtained and possibly even 3D surface representation of the vane. Numerical values obtained and views these images provide enough information to see trends and take appropriate corrective decisions for load range.

This impeller design phase involves many calculations and graphics based on large amounts of data. Optimum solution is possible using a structured computer program.

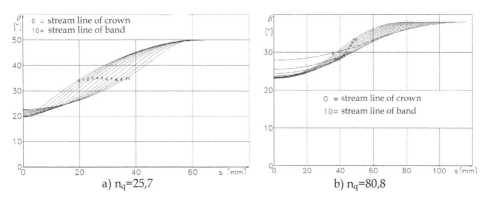

a) n_q=25,7   b) n_q=80,8

Fig. 16. Variation of β angle optimized for the two impeller types, (Milos, 2009)

11.2 Skeleton surface of the blade in projection on a plane perpendicular to the axis of rotation

Design studies and calculations of the impeller must be completed with an execution drawing. From the set up at this stage, the impeller blade contour was obtained in meridian plan. Starting from this design and adding some thick walls on the perimeter of front and rear shroud, followed by setting up the hub and impeller mounting area on the shaft, a drawing is obtained as a section through the impeller intersected by a plane passing through the rotation axis. Since the section is a piece generated by rotating, other representations are obtained easily in view. What remains not yet clearly defined are the 3D blades shape. The first step in this direction is to obtain the projection surface of the blade in a plane perpendicular to the axis of rotation.

In Figure 17, left, arc 1-2 is a streamline from the meridian plane. On this streamline at the current curvilinear coordinate x is considered discrete portion Δx. For this portion the projection image on a plane perpendicular to the axis of rotation is obtained. Streamline 1-2, when rotated around the axis of rotation generates surface flow on which moves the fluid particle driven by impeller blade. To highlight the trajectory that the blade must induce on a fluid particle on the strip surface flow generated by the arc Δx, is preferable that the strip is brought in the plane by developable surface. But the development of a curved surface geometry does not provide the necessary geometric links, so it is recourse to the projection of the strip surface flow on a conical surface, coaxial with the surface flow, which passes through the arc ends Δx. The error committed by the projection is relatively small given of the small size of the arc Δx. Basically bowstring Δx generates the conical surface in question. Developing the conical surface is performed immediately by arcs with the centre at cone top.

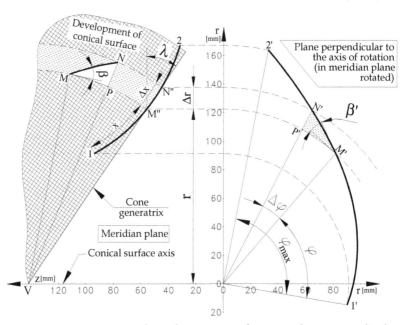

Fig. 17. Transposition route in-out from the stream surface on a plane perpendicular to the axis of rotation, (Milos, 2009)

Now on the strip a fluid particle trajectory can be traced by choosing a start point M and leading the segment MN at angle β against the tangent in M to the circle of the developed cone. The angle β corresponds to the average area of the chart Δx previously generated $\beta = f(x)$ and ultimately reflected in the average load on the blade portion Δx. Taking radius NV point P is obtained and the triangle MNP, with approximation, can be considered to be rectangular. Thus a sequence of fluid particle trajectory is obtained, needed to be done only if this route will provide a solid wall of the impeller blade.

Next is donne the transposition of this portion Δx from the meridian plane, on the flow surface, from developing the strip on the flow surface (from all three) in projection on a plane perpendicular to the axis of rotation. The plane in Figure 17 is the plane that was originally is perpendicular to rotation axis Oz, and then was rotated by 90° in the figure plane. This plane is in polar representation. The vertical axis is the axis Or crossing the meridian plan for this plane. Points 1 and 2 (beginning and end of the streamline of the blade) will be somewhere on the arcs of the same radius from the projection plane. Arc Δx in the projection will be between arcs of radius r and $r + \Delta r$. Let φ be polar angle measured from some reference straight. Blade portion Δx of the projection will appears under $\Delta\varphi$ angle. The following relations can be written:

a.   in developed conical surface passing through the ends of the arc Δx

$$tg\beta = \frac{NP}{MP}, \quad NP = \Delta x \tag{45}$$

b.   in the projection plane perpendicular to the axis of rotation

$$tg\beta' = \frac{N'P'}{M'P'} = \frac{\Delta r}{r\Delta\varphi}, \quad N'P' = \Delta r \tag{46}$$

Arc MP is projected in real size, becoming M'P ', i.e.

$$MP = M'P' = r\Delta\varphi \tag{47}$$

Replacing the first two relations we have:

$$tg\beta = \frac{\Delta x}{r\Delta\varphi} \tag{48}$$

$$tg\beta' = tg\beta\frac{\Delta r}{\Delta x} = tg\beta \cdot \cos\lambda \tag{49}$$

If you go to the limit with small infinites: $\Delta x \to dx$ and $\Delta\varphi \to d\varphi$ resulting differential equation:

$$tg\beta = \frac{dx}{r \cdot d\varphi} \tag{50}$$

Separating variables and integrating

$$\varphi = \int_{x_1}^{x} \frac{dx}{r \cdot tg\beta} \tag{51}$$

Integration is done along the streamline. The image in projection on a plane perpendicular to the axis of rotation is calculated and represents in the polar coordinates (r, φ), φ resulting in radians from the relation (51). To solve analytically the integral (51) there should be an analytical dependence $r = r(x)$ and $tg\beta = f(x)$. Since often discrete values of these dependencies are available, a numerical integration method is preferred. Integration is done by adding the number of partial areas using trapezoids (Fig. 12).

$$\Delta\varphi_i = \left[\left(\frac{1}{r \cdot tg\beta}\right)_i + \left(\frac{1}{r \cdot tg\beta}\right)_{i+1}\right]\frac{x_{i+1} - x_i}{2} \tag{52}$$

$\Delta\varphi_i$ angles calculated with (52) result in radians. Adding them step by step, current wrapping angles of the middle surface of blade are obtained.

$$\varphi_i = \sum_{i=1}^{i} \Delta\varphi_i \tag{53}$$

Thus each point (r_i, z_i) of the streamline is associated with an angle φ_i and results the defining of the middle surface of the blade in cylindrical coordinates (r_i, φ_i, z_i)

From hydrodynamic field calculations we have 3 ... 11 or even more streamlines. Applying relation (51) for each streamline will result φ_{max}. With φ_{max} you can start calculating φ optimization which seeks to obtain all streamlines wrapping up of the same angle range. Choosing as reference φ_{max}, max angle φ corresponding streamline medium (line 5 in case we have 11 streamlines) using a special algorithm changes the position of points x_3, position number between 1 and 99, until maximum φ is the same for each streamline. For the impellers with mixed-flow shape this may not be achieved. In this case it is necessary be provided that the differences between φ_{max} are uniformly increasing or decreasing.

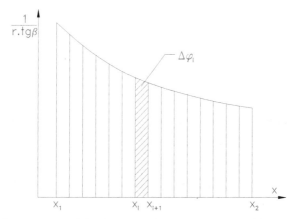

Fig. 18. Numerical integration of the function of wrapping angle of the blade

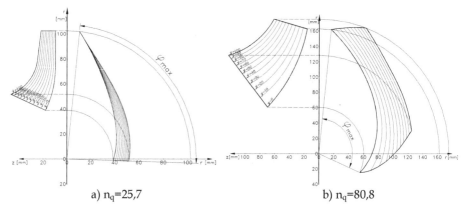

a) n_q=25,7   b) n_q=80,8

Fig. 19. Impeller blade in the meridian plane and in the projection plane, a plane perpendicular to the axis of rotation, (Milos, 2009)

With the interpolated angle β the middle blades projections in a plane perpendicular to the axis of rotation are calculated. The results are presented in Fig. 19. Figure 20 is presented as an example the projections of all impeller blades for the two distinct types of impellers with small n_q (radial) and large n_q (mixed-flow).

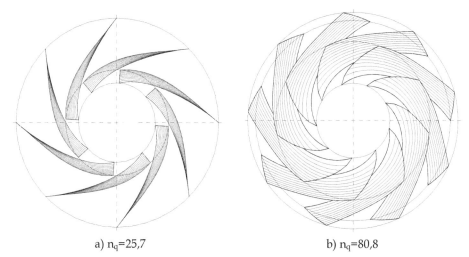

a) n_q=25,7   b) n_q=80,8

Fig. 20. Impeller blades in a plane perpendicular to the axis of rotation, (Milos, 2009)

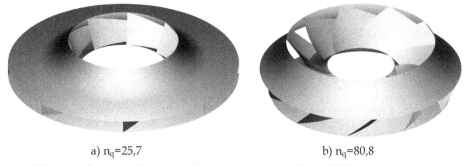

a) n_q=25,7   b) n_q=80,8

Fig. 21. 3D view of the impeller (only skeleton surfaces), (Milos, 2009)

12. Calculus of the skeleton blade surface in the space frame between inlet and outlet using conformal mapping method and CAD techniques

Regardless of the method of calculating the middle surface of the blade, the initial elements of construction are the angles of inlet, β_1 and outlet, β_2. Construction blade angles between inlet and outlet will be between these limits having a continuous and uniform variation on this course. In terms of geometric transformations **Conformal Mapping Method** (CMM) (Stepanoff, 1957) means flattening the trajectory on multiple conical surfaces, and in mathematical terms sets a bi-univocal correspondence between the middle curve of the blade on the stream surface and image plane of conformal transformation. Analytical transposition of this method is possible if we observe the conditions that must be fulfilled by function $A(x)$ from the image plane of conformal mapping (Fig. 23) correlated with the streamline of the meridian plane (Fig. 22). What this method adds to the classical ones is thickness accurate transposition versus the middle surface which is also rigorously calculated.

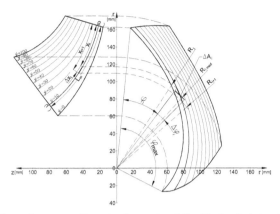

Fig. 22. Correspondence between discrete elements of the blade skeleton in the meridian plane and in projection onto a plane perpendicular to the axis of rotation

These conditions are expressed mathematically:

1. passing through point O: $A(0) = 0$;
2. passing through point M: $A(L_m) = A_M$;
3. angle of tangent to $A(x)$ in O to be identically with β_1: $\left.\dfrac{dA(x)}{dx}\right|_O = tg\,\beta_1$.

4. angle of tangent to $A(x)$ in M to be identically with β_2: $\left.\dfrac{dA(x)}{dx}\right|_M = tg\,\beta_2$;

Function which, as allure, is closest to the graphic-analytical solution is a third degree polynomial with four coefficients resulting from resolving the system of equations that is derived from applying the four conditions above.

Function $A(x) = f(x)$ polynomial of degree III, which defines the loading range between inlet and outlet and hence the middle surface geometry is:

$$A(x) = ax^3 + bx^2 + cx + d \tag{54}$$

Translating analytical the four conditions will result in a system of four equations with four unknowns (55) that solves exactly with Gauss elimination algorithm.

$$\begin{cases} A(x_O) = 0 & \text{(I)} \\ A(x_M) = A_M & \text{(II)} \\ A'(x_O) = tg\,\beta_1 & \text{(III)} \\ A'(x_M) = tg\,\beta_2 & \text{(IV)} \end{cases} \tag{55}$$

In Fig. 24 is graphically presented $A(x) = f(x)$ for the 11 streamlines of the meridian plane.

Basically, strips carried on the conical surface (Fig. 17) are here put together in the image plane. Function $A(x)$ must be at least level 3 to allow a second order derivative and thus be able to control the concavity on the work domain which must be uniformly increasing.

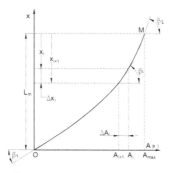

Fig. 23. Image plane of conformal mapping

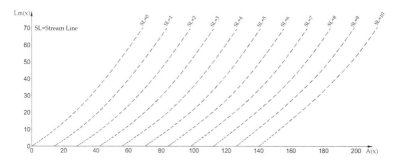

Fig. 24. Functions $A(x) = f(x)$ in a plane of conformal mapping for the 11 streamlines.

According to Figure 23 we see that the blade construction angle β_i results from:

$$tg\beta_i = \frac{\Delta x_i}{\Delta A_i} = \frac{x_{i+1} - x_i}{A_{i+1} - A_i} \tag{56}$$

According to notations in Fig. 22, elementary wrapping angle $\Delta\varphi_i$ of the projected blade result from the relationship:

$$\Delta\varphi_i = \frac{\Delta A_i}{\dfrac{R_i + R_{i+1}}{2}} = \frac{A_{i+1} - A_i}{\dfrac{R_i + R_{i+1}}{2}} \tag{57}$$

Current wrapping angle results from summation of elementary angles as calculated until the current angle by relationship:

$$\varphi_i = \sum \Delta\varphi_i \tag{58}$$

Using a special algorithm the same maximum wrapping angle of each streamline and desired maximum angle can be achieved so that the function $A(x)$ has no inflection and network profiles used in the next phase to be simple curved profiles. Difficulties in obtaining the same φ_{max} for all streamlines and their resolution are similar to those of the previous paragraph.

Once the middle surface of the blade is obtained, the next step is "dressing" this area with two adjacent surfaces that materializes every half-thickness of the blade. In the case of centrifugal pumps with small and medium n_q stream surface sections profiling is practiced with profiles of constant thickness and very low thickness (minimum 4 mm). Minimum thickness resulted from traditional technological conditions. If manufacturing technology allows and also the mechanical strength requirements are lower thickness can be lower when appropriate. The only adjustment is made at the inlet of the leading edge where rounding is practiced by a circle arc or ellipse arc, i.e. a semicircle or a semi-ellipse.

For the large and very large n_q, profiling is appropriate with shapes or profiles of least resistance to fluid movement. These can be obtained from catalogues or user-created profiles using the procedures known in hydrodynamics network profiles. In all cases (including constant thickness) only the thickness function applied to camber line arising from the CMM is used.

12.1 The transposition of *constant thickness* in the plane of conformal mapping (Milos, 2009)

Support curve is the streamline curve of conformal mapping plane. In a current point on the streamline curve of the conformal mapping plane having coordinates (A_i, L_{mi}) current angle β_i is already known. According to the schematic representation in Figure 25 the thickness range transposition relations in conformal mapping plane can be written.

-    for suction side

$$\begin{cases} A_{exi} = A_i + \dfrac{s}{2}\sin\beta_i \\ L_{exi} = Lm_i - \dfrac{s}{2}\cos\beta_i \end{cases} \tag{59}$$

-    for pressure side

$$\begin{cases} A_{ini} = A_i - \dfrac{s}{2}\cos\beta_i \\ L_{ini} = Lm_i + \dfrac{s}{2}\sin\beta_i \end{cases} \tag{60}$$

At inlet (leading edge) to avoid adverse effects on cavitation given by direct impacts of the fluid flow on a rough form, underside is connected with upside (or vice versa) by an arc of ellipse, Fig. 26. Given that this curve (ellipse) is treated in analytic geometry, those notations will be used.

If a and b are the semi-axes of the ellipse and s is the profile thickness of the leading edge, it appears that we always have $b = s/2$. Introducing the coefficient, $k_e = a/b$ it becomes a control parameter of the connection to leading edge. What must be determined in the first phase are discrete coordinates on the elliptical arc in relation to the local coordinate system $x'O'y'$. These in turn are determined by the mesh streamline curve (support) of the conformal mapping plan. These relations are:

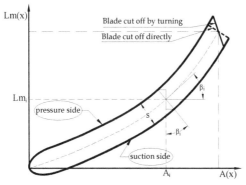

Fig. 25. Schematic representation of the thickness transposition in conformal mapping plan

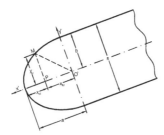

Fig. 26. Leading edge sides connected with an arc of ellipse

$$
\begin{cases}
x_M = k_e \dfrac{s}{2} - x_{cl} \\[4mm]
y_M = \pm \dfrac{1}{k_e} \sqrt{\left(k_e \dfrac{s}{2}\right)^2 - x_M}
\end{cases}
\tag{61}
$$

In subsequent calculations only enters y_M (with positive sign) which is semi-thickness with a positive sign and is put in the place of $s/2$ (59) and (60). For a good outline, especially in the front, it is necessary for a very fine mesh on the portion a. On the trailing edge the blade may be cut off straight or cut off by lathing without further processing. In the latter case we identify the extreme points.

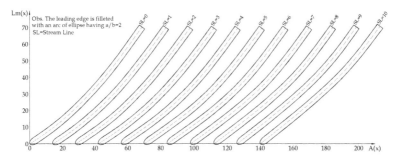

Fig. 27. Thickness transposition in conformal mapping plan for the 11 streamlines

For example shown in Fig. 27, results for a radial impeller thickness transposition studied in this respect.

12.2 The transposition of *variable thickness* in the plane of conformal mapping (Milos, 2009)

In this case most often are used known section thickness functions, preferably those for which an analytically given function of thickness is known. Support curve is the streamline curve of conformal mapping plane. In a current point on the curve of conformal mapping plane having coordinates (A_i, L_{mi}) angle β_i is already known. According to the schematic representation in Figure 28 the same blade thickness relations are used in terms of implementing the transformation line drawn from relations (59) and (60). Semi-variable thickness is calculated with a relation given or inferred.

For example in the case of NACA profiles the following relationship is used:

$$\frac{y_d}{l} = \frac{d}{l}\cdot\left[1{,}4845\cdot\sqrt{\frac{x}{l}}-0{,}63\cdot\frac{x}{l}-1{,}758\cdot\left(\frac{x}{l}\right)^2+1{,}4215\cdot\left(\frac{x}{l}\right)^3-0{,}575\cdot\left(\frac{x}{l}\right)^4\right] \qquad (62)$$

For calculations y_d is semi-thickness, and is put in the place of $s/2$ from relations (59) and (60). It must be noted that d in (62) is the maximum thickness of the profile and is required by the user. It may be constant or variable with average radius that is the flow surface, depending of resistance conditions imposed to mechanical strain or stiffness, vibration, etc. Using the results of thickness from 4-digit NACA profiles are presented as an example in Fig. 29.

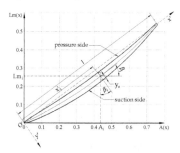

Fig. 28. Schematic representation of the variable thickness transposition of the profile in conformal mapping plane

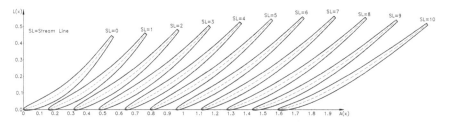

Fig. 29. Transposition of variable thickness in conformal mapping plane using 4-digit NACA profiles for the 11 streamlines

12.3 Determination of pressure surface and suction surface of the blade using the transposition of thickness in the plane of conformal mapping (Milos, 2009)

The process is similar to that of determining the middle surface of the blade except that in this case using a series of already known results for this surface. Having already solved the middle surface of the blade for any of the streamlines, and in conformal mapping planes having implemented thickness profiles (constant or variable) all the prerequisites are there for finding points and support curves of pressure and suction side surfaces. Start is made in conformal mapping plane where for a calculated point on the curve of the middle line of the blade, A_{sc}, is immediately founded L_{in} and L_{ex}, see route in Figure 30. Remember that values on axis $Lm(x)$ are the meridian streamline implementation on this axis in the plane of conformal mapping.

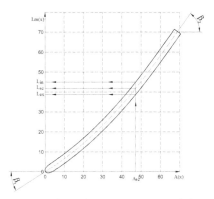

Fig. 30. Identification scheme in conformal mapping plane of the points from pressure surface and suction surface having the same wrapping angle φ in projection plane of the blade

The next step is to find by interpolation the position of these points on the streamline belonging to the meridian plane. Finding them depends on how you scroll the streamline. In the case of the optimization phase covering the camber line was recommended to go from outlet to the inlet. Now it is recommended covering to go from the inlet to the outlet, requiring a fine mesh at the leading edge, whether it is constant or variable thickness.

Points once found as positions along meridian plane, involves knowledge of pairs of coordinates (z, r). Passage in the projection plane (perpendicular to the axis of rotation) is made at the corresponding radiuses, followed by polar angle φ_{SC} (pole is the point of projection of the axis of rotation on this plane). Size A_{sc} contains, according to relation (57), the wrapping angle of middle line of blade, i.e. each A_i is assigned a φ_i. So as shown in Fig. 29, three points on the pressure surface, middle surface and suction surface are situated on the same polar straight at identical polar angle, φ_{sc}.

Here the three projection curves were represented schematically belonging to three 3D surfaces: pressure surface, middle surface and suction surface. In the cylindrical projection of the meridian plane they overlap. Resolving requires separate treatment for all their available streamlines. Adding the third coordinate φ, to the first two (z, r), completes the trio (z, r, φ), which is a representation in cylindrical coordinates. Based on the number of streamlines and their mesh representation results the smoothness of the surfaces determined in this way.

Note that the thickness of the profiles introduced by this method is the thickness measured on the surface flow. Real thickness (measured in the direction normal to local blade surface) is less or may be at most equal to that introduced in the conformal mapping plane. Therefore the strength and vibration calculations must take this into account.

With the given information the entire impeller and impeller form can be translated on a technical drawing. With the image from the meridian plane the impeller section and the impeller view are constructed. Projection on a plane perpendicular to the axis of rotation is only useful for the blade, which being of a more specific form, curved and twisted in space requires special methods of approach and drawing to be feasible technologically. The design itself consists of several designs resulted from blade cutting with two flat beams.

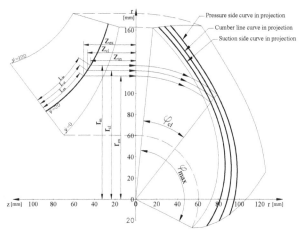

Fig. 31. Identification scheme, in meridian plane of the points from the pressure and suction surface followed by their projection in the plane perpendicular to the axis of rotation

The transition from cylindrical coordinate system in 3D Cartesian coordinate system is through relations:

$$\begin{cases} x = r \cdot \cos \varphi \\ y = r \cdot \sin \varphi \\ z = z \end{cases} \qquad (63)$$

In the case of impellers with high n_q, using this method, it is difficult to obtain a uniform blade surface which is generated of the curves determined by the conformal mapping transformation of each stream surface. This operation is done step by step, adjusting blade loading through angle β. Finally, if the wrapping angle range is not the same for every streamline or is not uniform ascending, the smoothing procedure is repeated.

3D representation of the results for visual examination is made in a CAD program with opportunities to read data from data files and their generation with smooth surfaces between the mesh networks. Fig. 32 presents a picture of a radial impeller obtained by CMM.

Fig. 32. 3D image of the rear shroud with blades (front shroud disk was removed)

13. Conclusion

Using these interpolation methods of β angle between inlet and outlet, optimized forms of the centrifugal pump blades can be obtained. What before, when the computer was not used, was approximated and required a great calculus effort, now it is possible to obtain in a few minutes. Starting with this way of solving the problem it is possible to imagine other interpolation functions with other restrictions.

The shape of the camber line resulted from this mode of interpolation of the angle β between inlet and outlet is very much alike with the one generated directly in the conformal transformation mapping plane, meaning that there are no inflexions and it is uniform increasing between inlet and outlet. This aspect suggests the possibility to couple the optimization of the blade shape through the classic method with the conformal transformation method.

The transposition of the thickness to the camber surface of the blade with the conformal transformation method gives this a very good precision compared to the classic method, and results in geometry of the blade that is much more improved.

14. Acknowledgment

The present work has been supported by the Romanian Government – Ministry of Education and Research, National Authority for Scientific Research through Research Grants, UEFISCDI department, project no. 21-036/2007 and project no. 21-41/2007.

15. References

Anton I., Câmpian V., Carte I., (1988), *Hidrodinamica turbinelor bulb şi a turbinelor pompe bulb*, Editura Tehnică, Bucureşti.

Anton L.E., Miloş T., (1998), *Centrifugal Pumps with Inducer*, Publishing house Orizonturi universitare, Timişoara, Romania, ISBN 978-973-625-838-1.

Gülich J.F., (2008), *Centrifugal Pumps*, Springer-Verlag Berlin Heidelberg New York, ISBN 978-3-540-73694-3.

Gyulai Fr., (1988), *Pumps, Fans, Compressors*; vol I & II, Publishing house of Politehnica University, Timişoara.

Miloş T., (2002), *Computer Aided Optimization of Vanes Shape for Centrifugal Pump Impellers*, Scientific Bulletin of The "Politehnica" University of Timişoara, Romania, Transactions on Mechanics, Tom 47(61), Fasc. 1, pp. 37-44, ISSN-1224-6077.

Miloş T., (2007), *Method to Smooth the 3d Surface of the Blades of Francis Turbine Runner*, International Review of Mechanical Engineering (IREME), Vol. 1, No. 6, pp. 603-607, PRAISE WORTHY PRIZE S.r.l, Publishing House, ISSN 1970-8734.

Miloş T., (2008), *CAD Procedure For Blade Design of Centrifugal Pump Impeller Using Conformal Mapping Method*, Fifth Conference of the Water-Power Engineering in Romania, Published in Scientific Bulletin of University POLITEHNICA of Bucharest, Series D, Mechanical Engineering, Vol.70/2008, No. 4, ISSN 1454-2358. pp. 213-220.

Miloş T., (2009), *Centrifugal and Axial Pumps and Fans*, Publishing house of Politehnica University, Timisoara, ISBN 978-973-625-838-1.

Miloş T., (2009), *Optimal Blade Design of Centrifugal Pump Impeller Using CAD Procedures and Conformal Mapping Method*, International Review of Mechanical Engineering (IREME), Vol. 3, No. 6, pp. 733-738, PRAISE WORTHY PRIZE S.r.l, Publishing House, ISSN 1970-8734.

Miloş T., Muntean S., Stuparu A., Baya A., Susan-Resiga R., (2006), *Automated Procedure for Design and 3D Numerical Analysis of the Flow Through Impellers*, In Proceedings of the 2nd German – Romanian Workshop on Vortex Dynamics, Stuttgart 10-14 May 2006. pp. 1-10. (on CD-ROM).

Pfleiderer K., (1961), *Die Kreiselpumpen für Flüssigkeiten und Gase*, Springer Verlag, Berlin.

Radha Krishna H.C. (Editor), (1997), *Hydraulic Design of Hydraulic Machinery*, Avebury, Ashghate Publishing Limited, ISBN: 0-29139-851-0.

Stepanoff A. J., (1957), *Centrifugal and Axial Flow Pumps*, 2nd edition, John Wiley and Sons, Inc., New York.

Tuzson J., (2000), *Centrifugal Pump Design*, John Wiley & Sons, ISBN 9780471361008.

Fault Diagnosis of Centrifugal Pumps Using Motor Electrical Signals

Parasuram P. Harihara[1] and Alexander G. Parlos[2]
[1]Corning Incorporated,
[2]Texas A&M University,
USA

1. Introduction

Centrifugal pumps are some of the most widely used pumps in the industry (Bachus & Custodio, 2003) and many of them are driven by induction motors. Failure to either the induction motor or the centrifugal pump would result in an unscheduled shutdown leading to loss of production and subsequently loss of revenue. A lot of effort has been invested in detecting and diagnosing incipient faults in induction motors and centrifugal pumps through the analysis of vibration data, obtained using accelerometers installed in various locations on the motor-pump system. Fault detection schemes based on the analysis of process data, such as pressures, flow rates and temperatures have also been developed. In some cases, speed is used as an indicator for the degradation of the pump performance. All of the above mentioned schemes require sensors to be installed on the system that leads to an increase in overall system cost. Additional sensors need cabling, which also contributes towards increasing the system cost. These sensors have lower reliability, and hence fail more often than the system being monitored, thereby reducing the overall robustness of the system. In some cases it may be difficult to access the pump to install sensors. One such example is the case of submersible pumps wherein it is difficult to install or maintain sensors once the pump is underwater. To avoid the above-mentioned problems, the use of mechanical and/or process sensors has to be avoided to the extent possible.

Motor current signature analysis (MCSA) and electrical signal analysis (ESA) have been in use for some time (Benbouzid, 1998, Thomson, 1999) to estimate the condition of induction motors based on spectral analysis of the motor current and voltage waveforms. The use of motor electrical signals to diagnose centrifugal pump faults has started to gain prominence in the recent years. However, it would be more beneficial if the drive power system (motor-pump system) as a whole is monitored. The large costs associated with the resulting idle equipment and personnel due to a failure in either the motor or the pump can often be avoided if the degradation is detected in its early stages (McInroy & Legowski, 2001). Moreover, the downtime can be further reduced if the faulty component within the drive power system can be isolated thereby aiding the plant personnel to be better prepared with the spares and repair kits. Hence there is not only a strong need for cost-effective schemes to assess the "health" condition of the motor-pump system as a whole, but also a strong requirement for an efficient fault isolation algorithm to isolate the component within the motor-pump system that is faulty. The unique contribution of this work is that it uses only the motor electrical signals to detect and isolate faults in the motor and the pump. Moreover, it does not presume the existing "health" condition of either the motor or the pump and detects the degradation of the system from the current state.

2. Literature review

Most of the literature on fault detection of centrifugal pumps is based on techniques that require the measurement of either vibration or other process based signals. There are very few peer-reviewed publications that deal with non-invasive/non-intrusive techniques to diagnose faults in centrifugal pumps. Even fewer literatures are available on the isolation of faults between the pump and the motor driving the pump. In this chapter, only the publications that deal with detecting centrifugal pump faults using motor electrical signals are reviewed. In (Dister, 2003), the authors review the latest techniques that are used in pump diagnostics. Hardware and software algorithms required to make accurate assessment of the pump condition are also discussed. Lists of typical problems that develop in the pump along with the conventional methods of detection are presented. In (Siegler, 1994), the authors describe the development and application of signal processing routines for the detection of eroded impeller condition of a centrifugal pump found in submarines. Fault features are extracted from the power spectrum and a neural networks-based classification scheme based on the nearest neighborhood technique classifies about 90%of the test cases correctly. In (Casada, 1994, 1995, 1996a) and (Casada & Bunch, 1996b), motor current and power analysis is used to detect some operational and structural problems such as clogged suction strainer and equipment misalignment. Load related peaks from the power or current spectrum are used as fault indicators in the proposed scheme. A comparative study between the vibration spectrum-based, power spectrum-based and torque spectrum-based detection methods is also described in detail. The authors conclude that the motor-monitored parameters are much more sensitive than the vibration transducers in detecting effects of unsteady process conditions resulting from both system and process specific sources. In (Kenull et al., 1997), the energy content of the motor current signal in specific frequency ranges are used as fault indicators to detect faults that occur in centrifugal pumps, namely, partial flow operation, cavitation, reverse rotation, etc. The work in (Dalton & Patton, 1998) deals with the development of a multi-model fault diagnosis system of an industrial pumping system. Two different approaches to model-based fault detection are outlined based on observers and parameter estimation. In (Perovic, Unsworth & Higham, 2001), fault signatures are extracted from the motor current spectrum by relating the spectral features to the individual faults to detect cavitation, blockage and damaged impeller condition. A fuzzy logic system is also developed to classify the three faults. The authors conclude that the probability of fault detection varies from 50% to 93%. The authors also conclude that adjustments to the rules or the membership functions are required so that differences in the pump design and operating flow regimes can be taken into consideration. In (Schmalz & Schuchmann, 2004), the spectral energy within the band of about 5 Hz to 25 Hz is calculated and is used to detect the presence of cavitation or low flow condition in centrifugal pumps. In (Welch et al., 2005) and (Haynes et al., 2002), the electrical signal analysis is extended to condition monitoring of aircraft fuel pumps. The front bearing wear of auxiliary pumps is selected to demonstrate the effectiveness of the proposed algorithm. The authors after considerable study establish that the best indicator of front bearing wear in the motor current spectrum is not any specific frequency peak but is the base or floor of the motor current spectrum. The noise floor of the current spectrum is observed to increase in all pumps having degraded front bearings. In (Kallesoe et al., 2006), a model-based approach using a combination of structural analysis, observer design and analytical redundancy relation (ARR) design is used to detect faults in centrifugal pumps driven by induction motors. Structural considerations are used to divide the system into two cascaded subsystems. The variables connecting the two subsystems are estimated using an adaptive observer and the fault detection is based on an ARR which is obtained using Groebner basis algorithm. The measurements used in the development of the fault detection method are the motor terminal voltages and currents and the pressure delivered by the pump. In (Harris et al., 2004), the authors describe a fault detection system for diagnosing potential pump system failures using fault features extracted from the motor current and the predetermined pump design parameters. In (Hernandez-Solis & Carlsson, 2010), the motor current and power signatures are analyzed to not only detect when cavitation in the pump is present, but also when it starts. The correlation between the pump cavitation phenomena and the motor power is studied at different pump operating conditions.

Most of the detection schemes presented in the above-mentioned literature are based on either tracking the variation of the characteristic fault frequency or computing the change in the energy content of the motor current in certain specific frequency bands. The characteristic fault frequency depends on the design parameters, which are not easily available. For example, the rolling element bearing fault frequency depends on the ball diameter, pitch, contact angle, etc (McInerny & Dai, 2003). This information is not available, unless the pump is dismantled. Changes in the energy content within certain specific frequency bands could also result due to changes in the power supply or changes in the load even without any fault in the pump or these changes could also occur if a fault initiates in the induction motor that is driving the pump. Hence, this would result in the generation of frequent false alarms. Based on these discussions it can be seen that there is a strong need to develop a non-intrusive/non-invasive fault detection and isolation algorithm to detect and isolate faults in centrifugal pumps that is not only independent of the motor and pump design parameters but also independent of power supply and load variations.

3. Overview of fault detection methods

A fault detection system is said to perform effectively if it exhibits a high probability of fault detection and a low probability of false alarms. Fig. 1 shows the different characteristics any fault detection method exhibits. If the detection scheme is too sensitive then it is likely to generate frequent false alarms which lead to operators questioning the effectiveness of the detection method. At the same time if the detection scheme is too insensitive then there is a chance of missing anomalies that might lead to a fault. Missed faults may lead to critical equipment failure leading to downtime. As a result a balance between the fault detection capability and the false alarm generation rate must be achieved when designing a fault detection scheme. The fault detection methods can be broadly classified into two broad categories, namely, signal-based fault detection methods and model-based fault detection methods. Fig. 2 compares the procedure of a signal-based and model-based fault detection method.

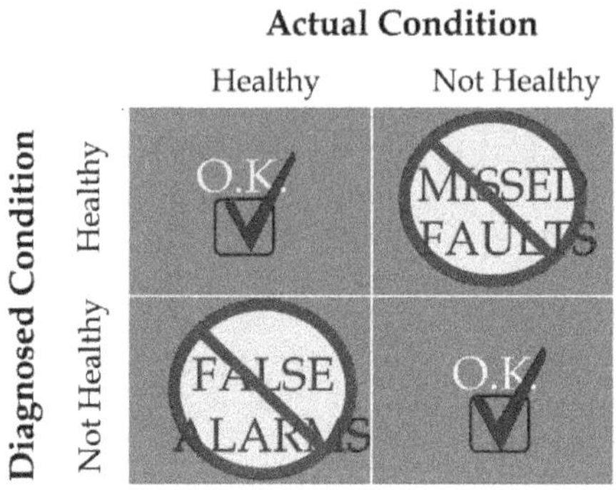

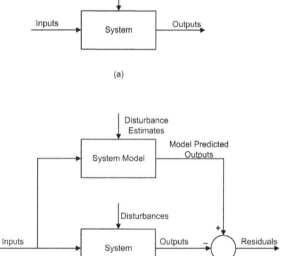

Fig. 1. Fault detection method characteristics

Fig. 2. (a) Signal-based fault detection method; (b) Model-based fault detection method

3.1 Signal-based fault detection method

Signal-based fault detection techniques are based on processing and analyzing raw system measurements such as motor currents, vibration signals and/or other process-based signals. No explicit system model is used in these techniques. Fault features are extracted from the sampled signals and analyzed for the presence or lack of a fault. However, these system signals are impacted by changes in the operating conditions that are caused due to changes in the system inputs and/or disturbances. Hence, if one were to analyze only the system signals for the presence of a fault, then it would be difficult to distinguish the fault related features from the input and disturbance induced features. This would result in the generation of frequent false alarms, which would in turn result in the plant personnel losing confidence over the fault detection method. If the system is considered to be ideal, i.e., there are no changes in the input and a constant input is supplied to the system and there are no disturbances affecting the system, then the signal-based detection schemes can be used in the detection of system faults with 0% false alarms. However, in reality such a system does not exist. The input variations cannot be controlled and harmonics are injected into the system and into the system signals. Moreover, disturbances to the system always occur and are always never constant. Hence these variations affect the system signals and result in the generation of false alarms.

3.2 Model-based fault detection method

The basic principle of a model-based fault detection scheme is to generate residuals that are defined as the differences between the measured and the model predicted outputs. The system model could be a first principles-based physics model or an empirical model of the actual system being monitored. The model defines the relationship between the system outputs, system faults, system disturbances and system inputs. Ideally, the residuals that are generated are only affected by the system faults and are not affected by any changes in the operating conditions due to changes in the system inputs and/or disturbances. That is, the residuals are only sensitive to faults while being insensitive to system input or disturbance changes (Patton & Chen, 1992). If the system is "healthy", then the residuals would be approximated by white noise. Any deviations of the residuals from the white noise behavior could be interpreted as a fault in the system.

In (Harihara et al., 2003), signal-based and model-based fault detection schemes are compared to a flip-of-a-coin detector as applied to induction motor fault detection. The results of the study can be extended to centrifugal pump detection also. Receiver operating characteristic (ROC) curves are plotted for all the three types of detection schemes and their performances are compared with respect to the probability of false alarms and probability of fault detection. For false alarm rates of less than 50%, the flip-of-a-coin fault detector outperformed the signal-based fault detection scheme for the cases under consideration. It was possible to achieve 100% fault detection capability using the signal-based detection method, but at the same time there was a very high probability of false alarms (about 50%). On the contrary, the model-based fault detection method operated with 0% false alarm rates and had approximately 89% fault detection capability. If the constraint on the false alarm probability was relaxed to about 10% then it was possible to achieve 100% fault detection capability using the model-based detection technique.

4. Proposed fault diagnosis method

The fault diagnosis algorithm can be broadly classified into a three-step process; namely, fault detection, fault isolation and fault identification. The proposed fault diagnosis method in this chapter addresses the first two steps of the diagnostic process. It combines elements from both the signal-based and model-based diagnostic approaches. An overall architecture of the proposed method is shown in Fig. 3.

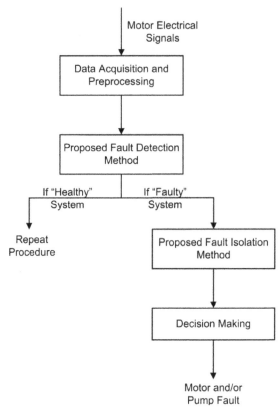

Fig. 3. Overall architecture of the proposed fault diagnosis method

The data acquisition module samples the three-phase voltages and three-phase currents. The data preprocessing module consists of down-sampling, scaling and signal segmentation. The sampled signals are down-sampled to match the sampling rate of the developed system model and normalized with respect to the motor nameplate information. In general, the motor electrical measurements are non-stationary in nature. However, traditional signal processing techniques such as FFT can be used to analyze these signals if quasi-stationary regions within these signals are identified. If identified, then only these segments of the signals are analyzed for the presence of a fault. A signal segmentation algorithm developed in this research is applied to the scaled motor electrical signals to determine the quasi-stationary segments within the signals. For a signal to be considered quasi-stationary, its fundamental frequency component and the corresponding harmonic components must remain constant over time. Thus as part of the signal segmentation algorithm, the time variations of the spectral components of the sampled signals are investigated and only those time segments of the sampled signals during which the spectral components are constant are considered for further analysis. Moreover, only the spectral components with large magnitudes are considered as those with very small amplitudes do not contribute significantly to the overall characteristics of the signal. Since the resulting signals are quasi-stationary in nature, Fourier-based methods can be applied to extract the fault features.

4.1 Proposed fault detection method

The schematic of the proposed fault detection method is shown in Fig. 4. As mentioned in the previous section, the proposed method combines elements from both the signal-based and model-based fault detection methods. The quasi-stationary segments of the pre-processed signals are used as inputs to both the "system model" module and the "fault feature extraction" module. Residuals are generated between the fault indicators extracted from the system signals and the fault features estimated by the system model. These residuals are further analyzed to detect the presence of a fault in the system.

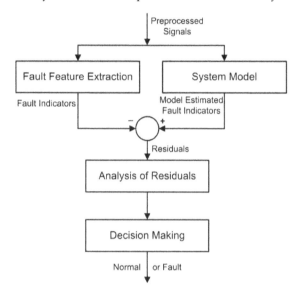

Fig. 4. Schematic of the proposed fault detection method

4.1.1 Description of the fault detection indicator

Most of the available literatures are based on extracting and tracking the variation of specific characteristic frequencies. There are certain limitations associated with this approach. One is the motor and/or pump design parameters or physical model parameters are required to obtain such characteristic frequencies. Secondly, the motor current spectrum is usually contaminated by load variations resulting in false indications of fault presence, though load compensation can remedy this. To overcome these limitations, the proposed fault indicator is based on monitoring the harmonic content of the motor current signals. This is based on the premise that any change in the ``health'' of the system would induce harmonic changes in the motor torque which would in turn induce harmonic changes in the motor current.

The Short Term Fourier Transforms (STFT) is used to process the motor current signals. In this study, the proposed fault indicator is defined as:

$$FDI(k) = \frac{1}{3} \sum_{a,b,c} \frac{\sum_k I_k^2}{I_f^2}. \tag{1}$$

where a, b and c are the three phases of the motor current, I_k is the RMS value of the kth harmonic component in the motor current and I_f is the fundamental frequency component of the motor current.

4.1.2 Description of the system model

To reduce the generation of false alarms and maintain a good fault detection capability, the effects of the changing input conditions must be isolated. In this study, this is accomplished by means of an empirical model. The developed model describes the relation between the baseline (or "healthy") response of the system and the system inputs. The baseline response of the system is described by the fault indicator of a "healthy" system. The inputs to the model are derived from the preprocessed system signals. They include energy content and harmonic distortion of the voltage signal, system load level etc. The model structure used in this study is of the form:

$$\Gamma\left[\Lambda\left(V(t)\right), \Psi\left(I(t)\right), FDI\right] = 0 \tag{3}$$

where Γ is the unknown function to be modeled, Λ is the transformation function that converts the preprocessed voltage signals to the system model inputs, Ψ is the transformation function that converts the preprocessed current signals to the system model inputs, $V(t)$ is the time varying preprocessed voltage signals, $I(t)$ is the time varying preprocessed current signals and FDI is the fault indicator described in the previous subsection. In this study, the unknown function Γ is modeled as a polynomial having the structure similar to a nonlinear ARX model. The accuracy of the model output depends on the nature (accuracy, volume, etc) of the raw data used in the training phase. Hence the system is operated in a sufficiently wide range to cover the entire operating envelope of interest. The proposed model is developed using data collected from the "healthy" baseline system. The developed model predicts the baseline fault indicator estimate for a given operating condition characterized by the model inputs. The model is validated using data that are different from the one used in its development.

Another important observation to note is that no fault data are used to train the model. Hence for anomalies in the pump or motor, the output of the model will be the system baseline fault indicator estimate (or the "healthy" system FDI estimate) for the given operating condition. No motor or pump design parameters are used in the development of the baseline model. Hence this model can be easily ported to other motor-centrifugal pump systems, as only the measured motor voltages and currents are used in model development. However, each motor-centrifugal pump system will have a different baseline model, which can be adaptively developed using the measured motor electrical signals.

4.1.3 Analysis of residuals and decision making

An average of the model estimated output ("healthy" system FDI estimate) is compared to the average of the FDI extracted from the measured signals and the residuals between the two are computed. The computed residual is then normalized with respect to the average of the model estimated output and is tracked over time. This normalized residual is defined as the fault detection indicator change (FDIC). Let the size of the moving window within the time segment $[t_1, t_N]$ be $(t_2 - t_1)$ and the moving distance of the window be p. The FDIC is computed as

$$FDIC = \frac{\displaystyle\sum_{i=t_1+kp}^{t_2+kp} FDI(i) - \sum_{i=t_1+kp}^{t_2+kp} F\hat{D}I(i)}{\displaystyle\sum_{i=t_1+kp}^{t_2+kp} F\hat{D}I(i)}, \quad k = 0, 1, 2, ...m \tag{4}$$

where $m = (t_N - t_2)/p$. If the system is "healthy", then the FDIC can be approximated by white noise. However, if there is a fault in the system, then the FDIC will deviate from the white noise behavior. If this deviation exceeds a certain threshold then a "fault" alarm is issued. Otherwise, the system is considered "healthy" and the procedure is repeated. If the detection threshold is chosen to be very large, then although the false alarm rates are reduced, there is a very high probability of missing a fault. Similarly, if the detection threshold is chosen to be very small then along with good fault detection capability, there is a very high probability of generating false alarms. Hence a balance has to be achieved in deciding the detection threshold. One factor in choosing the threshold is the intended application of the detection method or the system that is being monitored. For example, in space applications, a high rate of false alarms is acceptable as people's lives are at stake. Hence the threshold can be chosen small to detect any anomaly. In utility industries however, false alarms are not tolerated and hence a somewhat higher threshold is preferred. The detection method might not detect the fault as soon as the fault initiates, but will detect it as the fault degrades and well before any catastrophic failure. In this study, an integer multiple of the standard deviation of the "healthy" baseline variation is used as the detection threshold.

4.2 Proposed fault isolation method

The output of the system model developed in the previous subsection is affected by either a fault in the induction motor or a fault in the centrifugal pump or any other component affecting the motor output. The reason is that the model is developed for the entire system (motor-pump) as a whole. For the purpose of this study only motor and pump faults are assumed. Hence, it is not possible to isolate a developing fault. To distinguish between faults in the motor and faults in the pump, a localized model of one of the components is required wherein the output of the model is affected only by the faults in that component and is insensitive to the faults in the other. In this study, since no measurement is available from the centrifugal pump, a localized model for the induction motor is developed. The output of this model is only sensitive to faults in the motor and is insensitive to faults in the centrifugal pump. The fault isolation method is used to distinguish between motor and pump faults only when a fault within the system is detected. If the system is "healthy", then the next data set is analyzed to check for the presence or lack of fault and the fault isolation method is not used.

4.2.1 Development of the localized induction motor model

Consider an induction machine such that the stator windings are identical, sinusoidally distributed windings, displaced by 120°, with N_s equivalent turns and resistance, r_s.

Consider the rotor windings as three identical sinusoidally distributed windings displaced by 120°, with N_r equivalent turns and resistance, r_r. The voltage equations are given as:

$$v_{abcs} = r_s i_{abcs} + p\lambda_{abcs}$$
$$v_{abcr} = r_r i_{abcr} + p\lambda_{abcr} \tag{5}$$

where, v is the voltage, i is the current, λ is the flux linkage, p is the first derivative operator, subscript s denotes variables and parameters associate with stator circuits and subscript r denotes the variables and parameters associated with the rotor circuits. a, b and c represent the three phases. r_s and r_r are diagonal matrices each with equivalent nonzero elements and

$$\left(f_{abcs}\right)^T = \begin{bmatrix} f_{as} & f_{bs} & f_{cs} \end{bmatrix}$$
$$\left(f_{abcr}\right)^T = \begin{bmatrix} f_{ar} & f_{br} & f_{cr} \end{bmatrix} \tag{6}$$

where f represents either voltage, current or flux linkages. For a magnetically linear system, the flux linkages may be expressed as

$$\begin{bmatrix} \lambda_{abcs} \\ \lambda_{abcr} \end{bmatrix} = \begin{bmatrix} L_s & L_{sr}\left(\theta_m(t)\right) \\ L_{sr}^T\left(\theta_m(t)\right) & L_r \end{bmatrix} \begin{bmatrix} i_{abcs} \\ i_{abcr} \end{bmatrix} \tag{7}$$

where L_s and L_r are the winding inductances which include the leakage and magnetizing inductances of the stator and rotor windings, respectively. The inductance L_{sr} is the amplitude of the mutual inductances between the stator and rotor windings. L_s and L_r are constants and L_{sr} is a function of the mechanical rotor position, $\theta_m(t)$. Details of the variables are described in (Krause et al., 1994).

The vast majority of induction motors used today are singly excited, wherein electric power is transformed to or from the motor through the stator circuits with the rotor windings short-circuited. Moreover, a vast majority of single-fed machines are of the squirrel-cage rotor type. For a squirrel cage induction motor, $v_{abcr} = 0$. Substituting equation (7) in equation (5),

$$v_{abcs} = r_s i_{abcs} + L_s\left(p i_{abcs}\right) + \left(p L_{sr}\left(\theta_m(t)\right)\right) i_{abcr} + L_{sr}\left(\theta_m(t)\right)\left(p i_{abcr}\right)$$
$$0 = r_r i_{abcr} + \left(p L_{sr}^T\left(\theta_m(t)\right)\right) i_{abcs} + L_{sr}^T\left(\theta_m(t)\right)\left(p i_{abcs}\right) + L_r\left(p i_{abcr}\right) \tag{8}$$

In considering the steady state form of equation (8) we are mixing the frequency and time domain formulations for the sake of simplicity. Adhering to strict frequency or time domain representations provides the same qualitative results but it complicates the equations. The following steady state representation of equation (8) is obtained:

$$\tilde{V}_s(t) = \left(r_s + j\omega_s L_s\right)\tilde{I}_s(t) + \left(j\omega_s L_{sr}\left(\theta_m(t)\right)\right)\tilde{I}_r(t)$$
$$0 = j\omega_r L_{sr}^T\left(\theta_m(t)\right)\tilde{I}_s(t) + \left(r_r + j\omega_r L_r\right)\tilde{I}_r(t) \tag{9}$$

where, V_s is the stator voltage, I_s is the stator current, I_r is the rotor current and ω is the speed. In equation (9), assuming that $\left(r_r + j\omega_r L_r\right)$ is invertible, $\tilde{I}_r(t)$ can be expressed as

$$\tilde{I}_r(t) = -\frac{j\omega_r L_{sr}^T(\theta_m(t))}{r_r + j\omega_r L_r}\tilde{I}_s(t) \tag{10}$$

Substituting equation (10) in equation (9),

$$\tilde{V}_s(t) = \left(r_s + j\omega_s L_s + \frac{\omega_s \omega_r L_{sr}(\theta_m(t))L_{sr}^T(\theta_m(t))}{r_r + j\omega_r L_r}\right)\tilde{I}_s(t) \tag{11}$$

Assuming $\left(r_s + j\omega_s L_s + \dfrac{\omega_s \omega_r L_{sr}(\theta_m(t))L_{sr}^T(\theta_m(t))}{r_r + j\omega_r L_r}\right)$ is invertible,

$$\tilde{I}_s(t) = \left[\left(r_s + j\omega_s L_s + \frac{\omega_s \omega_r L_{sr}(\theta_m(t))L_{sr}^T(\theta_m(t))}{r_r + j\omega_r L_r}\right)\right]^{-1}\tilde{V}_s(t) \tag{12}$$

$$\tilde{I}_s(t) = \left[Z(\theta_m(t))\right]^{-1}\tilde{V}_s(t)$$

where Z is a function of the machine parameters which in turn are functions of the mechanical rotating angle of the rotor, $\theta_m(t)$. Equation (12) represents a modulator wherein the current spectrum will be composed of both the input voltage frequencies and also other frequency components due to the modulation. The modulated frequencies will appear as side-bands in the current spectrum around each frequency component corresponding to the input voltage signal. Hence an induction motor can be generalized as a modulator. Any fault in the rotor of the induction motor or in the motor bearings would result in the generation of additional spatial irregularities. This would induce additional spatial harmonics in the motor air-gap flux. These additional harmonics would modulate the voltage frequencies and appear as sidebands in the stator current spectrum. Higher order spectra are used to detect these modulated frequencies in the stator current spectrum.

4.2.2 Proposed fault isolation indicator

Higher-order spectra is a rapidly evolving signal processing area with growing applications in science and engineering. The power spectral density or the power spectrum of deterministic or stochastic processes is one of the most frequently used digital signal processing technique. The phase relationships between frequency components are suppressed in the power spectrum estimation techniques. The information contained in the power spectrum is essentially present in the autocorrelation sequence. This is sufficient for the complete statistical description of a Gaussian process of known mean. However, there are practical situations where the power spectrum or the autocorrelation domain is not sufficient to obtain information regarding deviations from Gaussian processes and the presence of nonlinearities in the system that generates the signals. Higher order spectra (also known as polyspectra), defined in terms of higher order cumulants of the process, do contain such information. In this study higher order spectra are used to detect the phase relationship between harmonic components that can be used to detect motor related faults. One of the most widely used methods in detecting phase coupling between harmonic components is the bispectrum estimation method. In fact, bispectrum is used in detecting and characterizing quadratic phase coupling.

Consider a discrete, stationary, zero-mean random process $x(n)$. The bispectrum of $x(n)$ is defined as

$$B(w_1, w_2) = \sum_{\tau_1 \to -\infty}^{\infty} \sum_{\tau_2 \to -\infty}^{\infty} c(\tau_1, \tau_2) e^{\left[-j(w_1 \tau_1 + w_2 \tau_2) \right]}$$

where

$$c(\tau_1, \tau_2) = E\left[x(n)x(n + \tau_1)x(n + \tau_2) \right] \tag{13}$$

where, $E[.]$ denotes the expectation operator. A class of technique called "direct" can be used to estimate the bispectrum. This technique uses the discrete Fourier transform (DFT) to compute the bispectrum as follows:

$$B(k_1, k_2) = E\left[X(k_1)X(k_2)X^*(k_1 + k_2) \right] \tag{14}$$

where $X(k)$ is the DFT of $x(n)$. From equation (14), it can be concluded that the bispectrum only accounts for phase couplings that are the sum of the individual frequency components. However, motor related faults manifest themselves as harmonics that modulate the fundamental frequency and appear as sidebands at frequencies given by $\left| f_e \pm mf_v \right|$, where f_e is the fundamental frequency and f_v is the fault frequency. Hence, the bispectrum estimate given by equation (14) detects only half of the coupling, as it does not detect the presence of the other half given by the difference of the two frequency components. Moreover, information about the modulation frequency has to be known to use this bispectrum estimate correctly. Hence to correctly identify the modulation relationship, a variation of the modified bispectrum estimator also referred to as the amplitude modulation detector (AMD) described in (Stack et al., 2004) is used.

The AMD is defined as:

$$A\hat{M}D(k_1, k_2) = E\left[X(k_1 + k_2)X(k_1 - k_2)X^*(k_1)X^*(k_1) \right] \tag{15}$$

From equation (15), it can be seen that both the sidebands of the modulation are accounted for in the definition. No information about the modulation frequency is utilized in computing the AMD. This is very useful since the motor related fault frequencies which modulate the supply frequency are very difficult to compute. These frequencies are dependent on the design parameters, which are not easily available. For example, the fault frequency pertaining to a motor rolling element bearing depends on the contact angle, the ball diameter, the pitch diameter, etc. Hence it is desirable to design an algorithm which does not require the motor design parameters. In this study, the AMD definition given by equation (15) is applied to the three phase motor current signals and to the three phase motor voltage signals to obtain the fault isolation indicator (FII).

4.2.3 Decision making

The average of the FII is computed and tracked over time. As mentioned in the previous subsection, since the FII is based on the model of the induction motor, it is only sensitive to faults that develop in the induction motor and insensitive to faults in the centrifugal pump. If a fault develops in the induction motor, spatial harmonics are generated that leads to the FII to increase over time as the fault severity increases. Hence if the FII increases beyond a threshold, then it can be concluded that the fault is in the motor and not the pump. At the same time, if a fault is detected and the FII does not increase over time, then it can be concluded that the fault is in the pump and not the motor. The determination of the threshold is similar to the procedure followed to determine the fault detection threshold described in the previous section.

5. Sample results

Various experiments in a laboratory environment were conducted to test and validate the detection and isolation capability of the developed method. Experiments were also conducted to test the number of false alarms that the method generates. In this chapter, results from a field trial and a sample result from the laboratory experiments are presented. For more details on the various laboratory experiments, refer to (Harihara & Parlos, 2008a, 2008b, 2010). The proposed fault detection and isolation method was applied in an industrial setup to monitor a boiler feed-water pump fed by a 400 hp induction motor. Since no specific motor and/or pump model or design parameters are used in the development of the algorithm, the algorithm could be easily scaled to the 400 hp motor-pump system. The induction motor is energized by constant frequency power supply and the motor electrical signals are sampled using the current transducers and voltage transducers that are standard installations. Fig. 5 shows an indicative time series plot of the per unit value of the sampled motor electrical signals and Fig. 6 shows the power spectral density of one of the line voltages and phase currents. As shown in Fig.6. it is very difficult to detect the presence of the fault just by inspecting the spectrum of the electrical signals. The sampled electrical signals are used as inputs to the proposed fault detection and isolation algorithm to determine the "health" of the system.

Fig. 7 shows the proposed $FDIC$ for the data sets from the power plant. Note that the FDI that is obtained from the sampled signals is not used for monitoring purposes because this might result in the generation of false alarms as described in the previous section. The FDI is always compared to the model prediction, $\hat{FDI}$ and only the relative change is used for monitoring purposes. Hence only the FDIC is shown in the figure for illustrative purposes. The motor electrical signals were sampled at different points of time within a 7 month period. After "Sampling Point 6", data was continuously sampled till the motor was shutdown. The first few data sets are used to develop the motor-pump system model. Once the model is developed the proposed fault detection method is used to monitor the "health" of the system. A load increase is detected and the designed method accounts for this load change and re-initializes the proposed FDIC. The developed algorithm detects the presence of a fault within the motor-pump system as evident by the FDIC exceeding the defined warning threshold. Once the fault is detected the data is used by the proposed fault isolation algorithm to identify which component within the system has developed the fault. Fig. 8 shows the FII over time. The first few data sets are used to model the induction motor and get a baseline response of the motor. Note that the FII increases over time even though the motor drawn current is constant. As mentioned in the previous section, since the FII is based on a model of the induction motor, it is only sensitive to faults in the motor and insensitive to faults in the pump. Since the FII increases over time, it can be concluded that the fault is indeed in the motor and not in the pump. The power plant performed a diagnosis of the motor after shutdown and found a fault in the motor bearing.

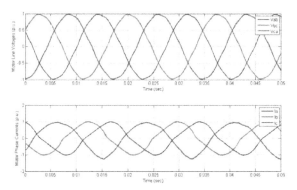

Fig. 5. Time series plot of the sampled motor electrical signals.

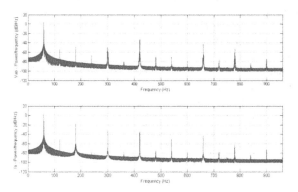

Fig. 6. Power spectral density of one of the line voltages and phase currents.

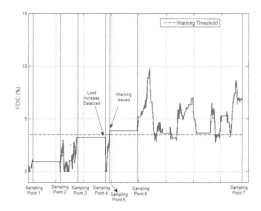

Fig. 7. Proposed fault detection method applied to data set from Texas A&M University Campus Power Plant detecting the presence of a fault in the motor-pump system.

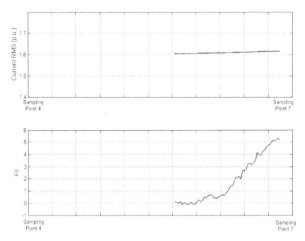

Fig. 8. Proposed fault isolation method applied to data set from Texas A&M University Campus Power Plant detecting the presence of a motor fault.

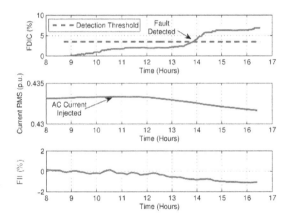

Fig. 9. Proposed fault detection and isolation method as applied to data set from a laboratory experiment; (top) proposed fault detection indicator change; (middle) motor current RMS; (bottom) proposed fault isolation indicator.

Fig. 9 shows the sample result from one of the laboratory experiments conducted to validate the performance of the proposed method on the detection and isolation of pump related failures. In this case study, one of the pump bearings is degraded using electric discharge machining (EDM) process. AC current of about 8A to 12 A is passed through the test bearing to accelerate the failure process. The top portion of the figure shows the FDIC which detects the fault immediately following the AC current injection. The middle portion of Fig.9 shows the change in the motor current. As the pump bearing is damaged the work output of the pump reduces which in turn results in the decrease of the input mechanical power. The decrease in the input power leads to a decrease in the motor current drawn. The bottom portion of the figure shows the FII based on the proposed method. As can be seen, the FII does not increase beyond the baseline variation since the developed model is insensitive to pump related faults and only sensitive to motor faults. This leads to the conclusion that the fault is indeed in the pump and not in the motor.

6. Summary

A novel fault detection and isolation method was proposed to detect and isolate centrifugal pump faults. The developed method uses only the motor electrical signals and is independent of the motor and/or pump design characteristics. Hence this method can be easily applied to other motor-pump systems. The proposed algorithm is also insensitive to power supply variations and does not presume the "health" condition of the motor or the pump. The developed fault detection and isolation method was applied in a field trail and was successful in detecting and isolating faults.

7. Acknowledgements

The research described in this chapter was conducted at Texas A&M University, College Station, TX, USA. The authors would like to acknowledge the financial support provided by the State of Texas Advanced Technology Program, Grants No. 999903-083, 999903-084, and 512-0225-2001, the US Department of Energy, Grant No. DE-FG07-98ID13641, the National Science Foundation Grants No. CMS-0100238 and CMS-0097719, and the US Department of Defense, Contract No. C04-00182.

8. References

Bachus, L. & Custodio, A. (2003). *Know and Understand Centrifugal Pumps*, Elsevier Advanced Technology, New York, USA.

Benbouzid, M. E. H. (1998). A Review of Induction Motors Signature Analysis as a Medium For Faults Detection, *Proceedings of the 24th Annual Conference of the IEEE Industrial Electronics Society*, pp. 1950-1955, ISBN 0 7803 4503 7, Aachen, Germany, Aug 31 – Sept 4, 1998.

Casada, D. A. & Bunch, S. L. (1996b). The Use of Motor as a Transducer to Monitor System Conditions, *Proceedings of the 50th Meeting of the Society for Machinery Failure Prevention Technology*, pp. 661-672, Jan, 1996.

Casada, D. A. (1994). Detection of Pump Degradation, *22nd Water Reactor Safety Information Meeting*, Bethesda, Maryland, USA, Oct 24-26, 1994.

Casada, D. A. (1995). The Use of Motor as a Transducer to Monitor Pump Conditions, *P/PM Technology Conference*, Indianapolis, Indiana, USA, Dec 6, 1995.

Casada, D. A. (1996a). Monitoring Pump and Compressor Performance Using Motor Data, *ASME International Pipeline Conference*, pp. 885-896, CODEN 002542, Calgary, Canada, Jun 9-13, 1996.

Dalton, T. & Patton, R. (19998). Model-Based Fault Diagnosis of a Two-Pump System, *Transactions of the Institute of Measurement and Control*, vol. 20, no. 3, (1998), pp. 115-124, ISSN 0142-3312.

Dister, C. J. (2003). Online Health Assessment of Integrated Pumps, *IEEE Aerospace Conference Proceedings,* pp. 3289-3294, ISBN 0-7803-7651-X, Big Sky, Montana, USA, Mar 8-15, 2003.

Harihara, P. P. & Parlos, A. G. (2008a). Sensorless Detection of Impeller Cracks in Motor Driven Centrifugal Pumps, *Proceedings of ASME International Mechanical Engineering Congress and Exposition,* pp. 17-23, ISBN 9780791848661, Boston, MA, USA, Oct 31 – Nov 6, 2008.

Harihara, P. P. & Parlos, A. G. (2008b). Sensorless Detection and Isolation of Faults in Motor-Pump Systems, *Proceedings of ASME International Mechanical Engineering Congress and Exposition,* pp. 43-50, ISBN 9780791848661, Boston, MA, USA, Oct 31 – Nov 6, 2008.

Harihara, P. P. & Parlos, A. G. (2010). Sensorless Detection of Cavitation in Centrifugal Pumps, *International Journal of COMADEM,* vol. 13, no. 2, (Apr 2010), pp. 27-33, ISSN 13637681.

Harihara, P. P., Kim, K. & Parlos, A. G. (2003). Signal-Based Versus Model-Based Fault Diagnosis – A Trade-Off in Complexity and Performance, *IEEE International Symposium on Diagnostics for Electric Machines, Power Electronics and Drives,* pp. 277-282, ISBN 0780378385, Atlanta, GA, USA, Aug 24-26, 2003.

Harris, C. A., Schibonski, J. A., Templeton, F. E. & Wheeler, D. L. (2004). Pump System Diagnosis, US Patent No: US 6,721,683 B2, Apr 2004.

Haynes, H. D., Cox, D. F. & Welch, D. E. (2002). Electrical Signature Analysis (ESA) as a Diagnostic Maintenance Technique for Detecting the High Consequence Fuel Pump Failure Modes, *Presented at Oak Ridge National Laboratory,* Oct 2002.

Hernandez-Solis, A. & Carlsson, F. (2010). Diagnosis of Submersible Centrifugal Pumps: A Motor Current and Power Signature Approaches, *EPE Journal,* vol. 20, no. 1, (Jan-March, 2010), pp. 58-64, ISSN 0939-8368.

Kallesoe, C. S., Cocquempot, V. & Izadi-Zamanabadi, R. (2006). Model-Based Fault Detection in a Centrifugal Pump Application, *IEEE Transactions on Control Systems Technology,* vol. 14, no. 2, (Mar 2006), pp. 204-215, ISSN 1063-6536.

Kenull, T., Kosyna, G. & Thamsen, P. U. (1997). Diagnostics of Submersible Motor Pumps by Non-Stationary Signals in Motor Current, *ASME Fluids Engineering Division Summer Meeting,* CODEN FEDSDL, Vancouver, Canada, Jun 22-26, 1997.

Krause, P. C., Wasynczuk, O. & Sudhoff, S. D. (1994). *Analysis of Electric Machinery,* Institute of Electrical and Electronics Engineers, ISBN 0780311019, New York, USA.

McInerny, S. A. & Dai, Y. (2003). Basic Vibration Signal Processing for Bearing Fault Detection, *IEEE Transactions on Education,* vol. 46, no. 1, (Feb 2003), pp. 149-156, ISSN 0018-9359.

McInroy, J. E. & Legowski, S. F. (2001). Using Power Measurements to Diagnose Degradations in Motor Drivepower Systems: A Case Study of Oilfield Pump Jacks, *IEEE Transactions on Industry Applications,* vol. 37, no. 6, (Nov/Dec 2001), pp. 1574-1581, ISSN 00939994.

Patton, R. J. & Chen, J. (1992). Robustness in Quantitative Model-Based Fault Diagnosis, *IEE Colloquium on Intelligent Fault Diagnosis – Part 2: Model Based Techniques,* pp. 4/1-4/17, Material Identity Number XX1992-00572, London, UK, Feb 26, 1992.

Perovic, S., Unsworth, P. J. & Higham, E. H. (2001). Fuzzy Logic System to Detect Pump Faults From Motor Current Spectra, *Proceedings of the 2001 IEEE Industry*

Applications Society 36th Annual Meeting, pp. 274-280, ISBN 0780371143, Chicago, IL, USA, Sept 30-Oct 4, 2001

Schmalz, S. C. & Schuchmann, R. P. (2004). Method and Apparatus of Detecting Low Flow/Cavitation in a Centrifugal Pump, US Patent No: US 6,709,240 B1, Mar 2004.

Siegler, J. A. (1994). Motor Current Signal Analysis for Diagnosis of Fault Conditions in Shipboard Equipment, *U.S.N.A Trident Scholar Project Report,* no. 220, U.S. Naval Academy, Annapolis, Maryland, USA, 1994.

Stack, J. R., Harley, R. G. & Habetler, T. G. (2004). An Amplitude Modulation Detector for Fault Diagnosis in Rolling Element Bearings, *IEEE Transactions on Industrial Electronics,* vol. 51, no. 5, (Oct 2004), pp. 1097-1102, ISBN 0278-0046.

Thomson, W. T. (1999). A Review of On-line Condition Monitoring Techniques For Three-Phase Squirrel Cage Induction Motors. Past Present and Future, *IEEE International Symposium on Diagnostics For Electrical Machines, Power Electronics and Drives,* pp. 3-18, ISBN 84 699 0977 0, Gijon, Spain, Sept 1-3, 1999.

Welch, D. E., Haynes, H. D., Cox, D. F., & Moses, R. J. (2005). Electric Fuel Pump Condition Monitor System Using Electrical Signature Analysis, US Patent No: US 6,941,785, Sept 2005.

4

Fluid Flow Control

Cristian Patrascioiu

Petroleum Gas University of Ploiesti,
Romania

1. Introduction

The present chapter is dedicated to the general presentation of the control system structures for the flow control in the hydraulic systems that have as components centrifugal pumps. The chapter also contains modeling elements for the following hydraulic systems: pumps, pipes and the other hydraulic resistances associated to the pipe.

1.1 The structure of the fluid flow control systems

The fluid flow control systems can be classified depending on the type of pressure source, on the pipe system structure and on the control element. Depending on the pressure source type, the flow control systems can be equipped with centrifugal pumps or volumetric pumps. Concerning the pipe structure, the flow control systems can be used within the hydraulic systems with branches or without branches. The control element within the flow control systems can be the control valve or the assembly variable frequency drive – electric engine - centrifugal pumps.

Throughout the next part there will be taken into consideration only the flow control systems having within their structure centrifugal pumps and pipe without branches. Due to this situation, the chapter will contain only two flow control systems:

a. The control system having the control valve as a control element, figure 1;
b. The control system having the assembly variable frequency drive – electric engine – centrifugal pump as a control element,, figure 2.

The flow control system having as control element the control valve consists of:

-   Process, made of centrifugal pump, pipe without brances, local hydraulic resistances;
-   Flow transducer, made of a diaphragm as primary element and a differential pressure transducer;
-   Feedback controller with proportional – integrator control algorithm;
-   Control valve.

The operation of the control system is based on the controlled action of the control valve hydraulic resistance, so that the hydraulic energy introduced by the centrifugal pump ensures the fluid circulation, within the flow conditions imposed and at the pressure of vessel destination and recovers the pressure loss associated to the pipe, associated to the local hydraulic resistances and associated to the control valve. The study and the design of this flow control system need the mathematical modeling of the centrifugal pump, of the pipe, of the local resistances, as well as of the control valve hydraulic resistance.

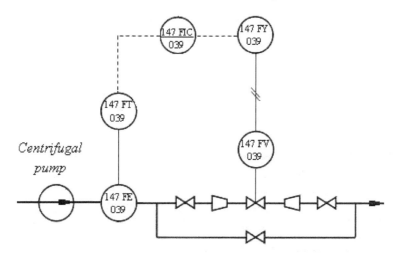

Fig. 1. The flow control system having as control element the control valve: FE – the sensitive element (diaphragm); FT – differential pressure transducer; FIC – flow controller; FY – electro-pneumatic convertor; FV – flow control valve.

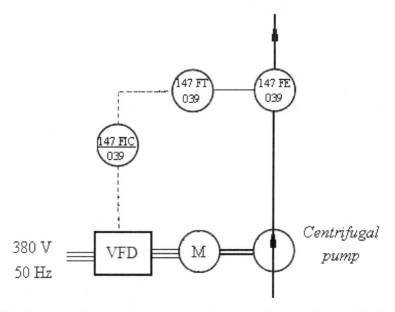

Fig. 2. The flow control system having as control element the assembly variable frequency drive – electric motor – centrifugal pump: M – electrical engine; VFD – variable frequency drive.

The flow control system having as control element the assembly - frequency static convertor – electric motor – centrifugal pump is made of:

- Process, made of a pipe without branches and local hydraulic resistances;
- Flow transducer, made of a primary element (diaphragm) and a differential pressure transducer;
- Feedback controller with proportional –integrator control algorithm;
- Control element made of variable frequency drive – electric motor – centrifugal pump.

The operation of this control flow system is based on the controlled actions of the centrifugal pump rotation, so that the hydraulic energy introduced ensures the fluid circulation, within the flow conditions imposed and the pressure in the destination vessel and recovers the pressure loss associated to the pipe and to the local hydraulic resistances. The study and the design of this control system needs the mathematical modeling of the centrifugal pump, of the pipe, of the local hydraulic resistances, as well as of the electric motor provided with frequency convertor.

1.2 The mathematical models of the fluid flow control elements

From the aspects presented so far there resulted the fact that the study and the design of the two types of control flow systems canot be done witout the mathematical modeling of all the subsystems that compose the control system. As a consequence, in the next part there will be presented: the mathematiocal model of the centrifugal pump, the mathematical model of the pipe, as well as of the local hydraulic resistance.

1.2.1 The simplified model of the centrifugal pumps

The operation parameters for the centrifugal pumps are: the volumetric flowrate, the ouput pumping pressure, the manometric aspiration pressure, the rotation speed, the hydraulic yield and the power consumption. The dependency between these variables is obtained experimentally. The obtained data is represented graphically, the diagrams obtained being named "the characteristics diagrams of the pump operation". In order to choose a centrifugal pumps to be used in a hydraulic system, there are applied the "characteristic diagrams at the constant rotation speed of pump".

The centrifugal pumps characteristics depend on the pumps constructive type. To exemplify, there has been chosen a pumps family used in refineries, figure 3 (Patrascioiu et al., 2009). The mathematical model of a centrifugal pump can be approximated by the relation

$$P_0 = a_0 + a_1 Q + a_2 Q^2. \tag{1}$$

Nine types of pumps have been selected out of the types and characteristics presented in figure 3. For each of these there has been extracted data regarding the flow (the independent variable) and the outlet pressure (the output variabile). The data has been processed by using the polynomial regression (Patrascioiu, 2005), the results obtained being presented in table 1. Based on the numerical results, there have been graphically represented the characteristics calculated with the relation (1) for the following type of pumps: the pump 32-13, the pump 50-20 and the pump 150-26, figure 4.

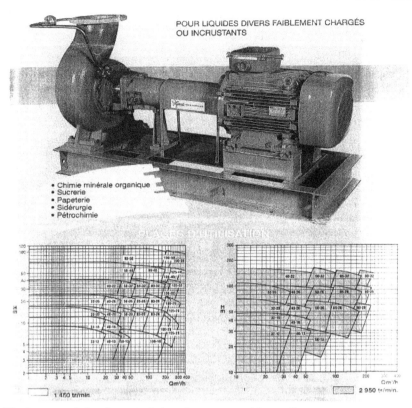

Fig. 3. The image and the characteristics of pumps used in a refinery

Pump type	Model coefficients (1)			Standard deviation [bar]
	a_0	a_1	a_2	
32-13	6.614213E+00	4.727929E-02	-3.938231E-02	6.621437E-02
32-16	8.851658E+00	7.501579E-01	-1.081358E-01	3.631092E-01
50-20	1.418321E+01	-2.746576E-02	-1.080953E-03	2.887783E-01
150-26	2.037071E+01	-7.850898E-03	-3.007388E-05	4.614905E-01
32-12	9.351883E+00	-1.262669E-01	3.939355E-04	1.890947E-01
50-13	8.227606E+00	-4.412075E-02	-1.679803E-05	1.578524E-01
40-16	1.258403E+01	-9.795947E-02	7.796262E-05	5.434285E-01
32-20	1.781198E+01	-2.755295E-01	1.361482E-03	3.952989E-01
40-20	1.742569E+01	-1.339487E-01	2.504382E-04	2.826216E-01

Table 1. The mathematical model coefficients of the pump family

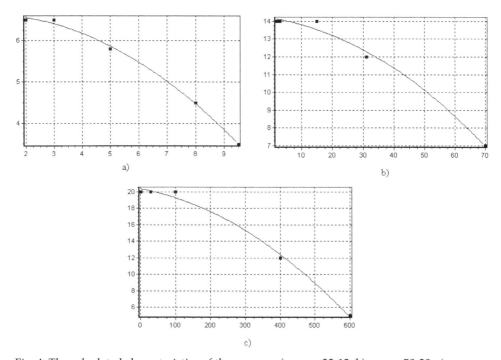

Fig. 4. The calculated characteristics of the pumps: a) pump 32-13; b) pump 50-20; c) pump 150-26.

1.2.2 The model of the pipe flow

The pipe represents a resistance of the hydraulic system. The mathematical model of the pressure lost by friction, for a straight pipe, with a circular section, is expressed by

$$\Delta P_{pipe} = \lambda \frac{8L}{\pi^2 D^5} Q^2 \quad \left[\frac{N}{m^2}\right] \tag{2}$$

where: λ – is the friction coefficient; L –the pipe length, in meters; D – the pipe diameter, in meters; Q – the fluid volumetric flow, in m³/s.

The application of the relation (2) implies determining the friction coefficient λ. The λ value depends mainly on the flowing regime, characterized by the Reynolds number Re, by the rugosity ε/D and the diameter D of the pipe. The Reynolds number is expressed by the relation

$$Re = \frac{Dw}{v}, \tag{3}$$

where w represents the fluid linear velocity; v- the fluid's dynamic viscosity.

For the calculus of the friction coefficient λ there are used the relations (Soare 1979):

$$\lambda = \begin{cases} \lambda = \dfrac{64}{\mathrm{Re}}, & \mathrm{Re} < 2300 \\[2mm] \dfrac{1}{\sqrt{\lambda}} = 1.74 - 2\lg\left(2\dfrac{\varepsilon}{D} + \dfrac{18.7}{\mathrm{Re}\sqrt{\lambda}}\right), & 2300 < \mathrm{Re} < 3000 \\[2mm] \dfrac{1}{\sqrt{\lambda}} = -2\lg\left(\dfrac{D}{3.7\varepsilon} + \dfrac{2.51}{\mathrm{Re}\sqrt{\lambda}}\right), & \mathrm{Re} > 3000 \end{cases} \tag{4}$$

The relations group (4) contains two nonlinear equations, their solution being the friction coefficient λ (Patrascioiu et al., 2009). Then, for the intermediate flow regime, $2300 < \mathrm{Re} < 3000$, is available this relation

$$\frac{1}{\sqrt{\lambda}} = 1.74 - 2\lg\left(\frac{2\varepsilon}{D} + \frac{18.7}{\mathrm{Re}\sqrt{\lambda}}\right).$$

This relation is brought to the expression of the nonlinear equation

$$f(\lambda) = 1.74 - 2\lg\left(\frac{2\varepsilon}{D} + \frac{18.7}{\mathrm{Re}\sqrt{\lambda}}\right) - \frac{1}{\sqrt{\lambda}} = 0. \tag{5}$$

For the turbulent flowing regime, $\mathrm{Re} > 3000$, the relation

$$\frac{1}{\sqrt{\lambda}} = -2\lg\left(\frac{\varepsilon}{3,7D} + \frac{2.51}{\mathrm{Re}\sqrt{\lambda}}\right)$$

represents a nonlinear equation

$$f(\lambda) = -2\lg\left(\frac{\varepsilon}{3.7D} + \frac{2.51}{\mathrm{Re}\sqrt{\lambda}}\right) - \frac{1}{\sqrt{\lambda}} = 0. \tag{6}$$

All nonlinear equations have been solved by using the numerical aghorithms (Patrascioiu 2005). The mathematical model of the pressure loss was simulated for the hydraulic system presented in table 2.

Variable	Measure unit	Value
Pipe		
Diameter	m	0.05
Length	m	20
Rugosity	-	0.03
Max flow rate	m³/h	
Fluid		
Viscosity	m² s⁻¹	0.92e-6
Density	Kg m⁻³	476

Table 2. The geometrical characteristics of the pipe and the physical properties of the fluid

The variation of the friction factor used within model (2) was calculated with the relations (4), the result being illustrated in figure 5. The increase of the fluid flow, its rate and the Reynolds factor respectively, leads to the decrease of the pipe-fluid friction factor.

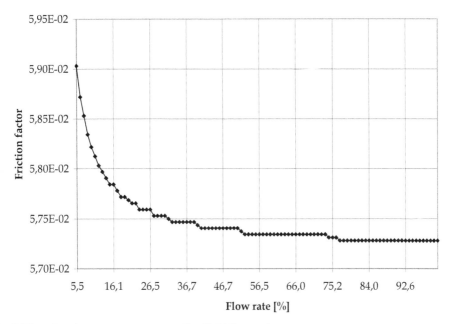

Fig. 5. The pipe drop pressure versus the fluid flow rate

In figure 6 there is presented the variation of the pressure drop in the pipe depending on the fluid flow. Due to the theoretical principles, the pressure drop on the pipe has a parabolic variation in relation to the fluid flow, although the friction factor decreases depending on the fluid rate.

1.2.3 The model of the hydraulic resistors

The local load losses, at the turbulent flow of a fluid by a restriction in the hydraulic system that modifies the fluid rate as size or direction are expressed either in terms of kinetic energy by relations under the form:

$$h_{hr} = \zeta \cdot \frac{w^2}{2g} \ , [m] \ , \tag{7}$$

or

$$\Delta p_{hr} = \zeta \cdot \frac{\rho w^2}{2} \ , [N/m^2] \ , \tag{8}$$

or in terms of linear load loss through an equivalent pipe of the length l_{hr} that determines the same hydraulic resistance as the considered local resistance

$$l_{hr} = \frac{\zeta}{\lambda} \cdot D .$$
(9)

The values of the local load loss coeffiecient ζ are usually obtained experimentally, in an analytical way, being estimated only in the case of turbulent flowing of a newtonian liquid. In table 3 there are presented the values of equivalent lengths (in metres) for different types of local resistances.

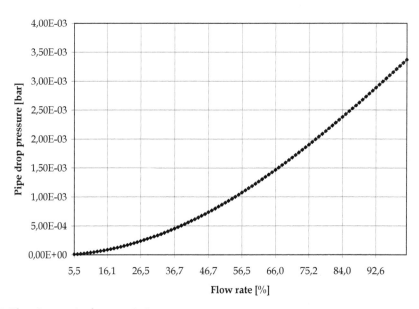

Fig. 6. The pipe static characteristic

The pipe nominal diameter [mm] / Local resistance	50	100	150	200	300	400	500
T-square	4.5	9.0	14.5	20.0	34.0	37.0	63.0
Crossover tee	5.0	11.5	17.5	26.0	47.0	74.0	100.0
Quarter bend: $\alpha = 90°$; $Re/R = 8$	1.0	1.7	2.5	3.2	5.0	7.0	9.0
Quarter bend: $\alpha = 90°$; $Re/R = 6$	1.5	2.5	4.0	5.0	7.5	11.0	44.0
Cast curve	3.2	7.5	12.5	18.0	30.0	44.0	55.0
Slide valve	0.6	1.5	2.0	3.0	5.0	7.5	10.0
Tap valve	0.6	-	1.2	1.8	-	-	-
Flat compensator of expansion shape	4.0	9.5	14.5	20.0	33.0	48.0	64.0
Choppy compensator of expansion	5.0	12.0	18.5	26.0	42.0	61.0	82.0
Safety valve	3.6	7.5	12.5	18.0	130.0	-	-
Valve with normal pass	13.0	31.0	50.0	73.0	130.0	200.0	270.0
Valve with bend pass	10.0	20.0	32.0	45.0	77.0	115.0	150.0

Table 3. The equivalent lengths (metres of pipe) of some local resistances

2. Flow transducers

The flow transducer is included within the structure of the automatic system of flow control. The design of the flow control systems includes the stage of choosing the transducer type and its sizing. From the author's experience, in the domain of the chemical engineering there are especially used flow transducers based on the fluid strangling. From these, the most representative ones are the flow transducers with a diaphragm (for pipes with a circular section) and the flow transducers with a spout with a long radius (for the rectangular flowing sections).

2.1 The flow transducers with a diaphragm

The decrease section method is governed by national standards (STAS 7347/1-83, 7347/2-83, 7347/3-83). Within the flow transducer, the primary element is the diaphragm, classified as follows:

- with pressure plugs an angle;
- with pressure plugs at D and $D/2$;
- with pressure plugs in flange.

Constructive elements specific to diaphragms are presented in figure 7.

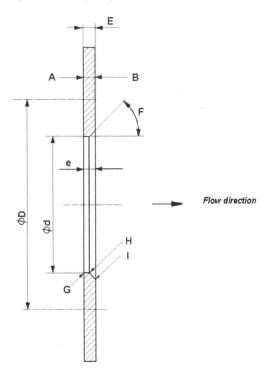

Fig. 7. The normal diaphragm construction: A – upstream face; B – downstream face; E – plate thickness; G – upstream edge; H,I – downstream edges; e – hole thickness.

The domain of the diaphragm is restricted to circular section pipes, the diameter of the pipe and of the diaphragm being also restricted, table 4.

Characteristic	Plugs at flange	Plugs at D and $D/2$	Plugs in angle
d [mm]	≥ 12.5	≥ 12.5	≥ 12.5
D [mm]	$50 \leq D \leq 760$	$50 \leq D \leq 760$	$50 \leq D \leq 1000$
β	$0.2 \leq \beta \leq 0.75$	$0.2 \leq \beta \leq 0.75$	$0.23 \leq \beta \leq 0.80$
Re_D	$\geq 1260 \beta^2 D \leq 10^8$	$\geq 1260 \beta^2 D \leq 10^8$	$5000 \leq Re_D \leq 10^8$ $0.23 \leq \beta \leq 0.45$ $10000 \leq Re_D \leq 10^8$ $0.45 \leq \beta \leq 0.77$ $20000 \leq Re_D \leq 10^8$ $0.77 \leq \beta \leq 0.80$

Table 4. Diaphragms use domain

The mass flow, Q_m , is calculated with the relation

$$Q_m = CE\varepsilon \frac{\pi}{4} d^2 \sqrt{2\, \Delta p\, \rho_1}\; . \; [\text{kg/s}] \tag{10}$$

The significance of the variables is:

-    C discharge coeffiecient,

$$C = \frac{\alpha}{E}\; ; \tag{11}$$

-    d diameter of the primary hole   [m];
-    D pipe diameter   [m];
-    β diameter ratio,

$$\beta = \frac{d}{D}\; ; \tag{12}$$

-    E closing rate coeffiecient,

$$E = \frac{1}{\sqrt{1 - \beta^4}} \tag{13}$$

-    Δp differential pressure   [Pa];
-    ε expansion coefficient
-    ρ_1 fluid density upstream the diaphragm.   [kg/m³]

The discharge coefficient is given by the Stoltz equation

$$C = 0.5959 + 0.0312\,\beta^{2.1} - 0.1840\,\beta^8 + 0.0029\,\beta^{2.5}\left[\frac{10^6}{Re_D}\right]^{0.75}$$

$$+0.0900\,L_1\,\beta^4\left(1-\beta^4\right)^{-1} - 0.0337\,L_2'\,\beta^3. \tag{14}$$

The volumetric flow, Q_v, is calculated with the classic relation

$$Q_v = \frac{Q_m}{\rho}.\ [\text{m}^3/\text{s}] \tag{15}$$

Within Stoltz equation, L_1, L_2 variables, respectively, are defined as follows:

- L_1 is the ratio between the distance of the upstream pressure plug measured from the diaphragm upstream face and the pipe diameter

$$L_1 = l_1 / D; \tag{16}$$

- L_2 is the ratio between the distance of the downstream pressure plug, measured by the diaphragm downstream face and the pipe diameter

$$L_2' = l_2' / D. \tag{17}$$

The particular calculus relations for L_1 şi L_2' are presented in table 5. The expansion coeffiecient ε is calculated irrespective of the pressure plug type, with the empirical relation

$$\varepsilon = 1 - \left(0.41 + 0.35\,\beta^4\right)\frac{\Delta p}{\chi\,p_1} \tag{18}$$

this relation being applicable for $\frac{p_2}{p_1} \ge 0.75$.

The pressure plugs type	Calculus relations	Observations
Pressure plugs in angle	$L_1 = L_2' = 0$	-
Plugs at D and D/2	$L_1 = 1$ $L_2' = 0.47$	As $L_1 \le 0.4333$, $\beta^4\left(1-\beta^4\right)^{-1} = 0.039$
Plugs at flange	$L_1 = L_2' = 25.4 / D$	For the pipes with the diameter $D \le 58.62$ mm, $L_1 \le 0.4333$, respectively $\beta^4\left(1-\beta^4\right)^{-1} = 0.039$

Table 5. Calculus relations for L_1 and L_2'

2.3 Case studies: The flow and diaphragm calculus for a flow metering system

In this paragraph there will be presented the calculus algorithm of the flow associated to a flow metering system and the diaphragm calculus algorithm used for the design of the metering system. The two algorithms are accompanied by industrial applications examples.

2.3.1 Algorithm for the calculus of the flow through diaphragm

The fluid flow that passes through a metering system having as primary element the diaphragm or the spout cannot be determined directly by evaluating the relation (10) due to the dependency of the discharge coefficient in ratio with the fluid rate, $C = f(v)$. Based on the relations presented in the previously mentioned standard, there has been elaborated a calculus algorithm of the fluid flow that passes through a metering system having the diaphragm as a sensitive element. Starting from the relation (10) there is constructed the nonlinear equation

$$g(Q_m) = 0, \tag{19}$$

where the function $g(Q_m)$ has the expression

$$g(Q_m) = Q_m - CE\frac{\pi}{4}d^2\sqrt{2\rho\Delta P}. \tag{20}$$

As the factors E, ε, d, ΔP and ρ do not depend on Q_m, the relation (20) can be expressed under the form:

$$g(Q_m) = Q_m - KC, \tag{21}$$

where

$$K = E\varepsilon\frac{\pi d^2}{4}\sqrt{2\rho\Delta P}. \tag{22}$$

Solving the equation (21) is possible, using the successive bisection algorithm combined with an algorithm for searching the interval where the equation solution is located (Patrascioiu 2005). Based on the algorithm presented, there was achieved a flow calculus program for a given metering system.

2.3.2 Industrial application concerning the flow calculus when diaphragm is used

A flow metering system is considered, having the following characteristics:

- Pipe diameter = 50 mm
- Diaphragm diameter = 35 mm
- Measuring domain of the differential pressure transducer = 2500 mmH$_2$O
- Fluid density = 797 kg/m^3
- Fluid viscosity = 3.76×10^{-6} m^2/s

The request is the determination of the flow valve corresponding to the maximum measured differential pressure.

The results of the calculus program contain the values of the discharge coefficient C, the closing-up rate coeffiecient E, the diameters ratio β, as well as the mass flow value Q_m, calculated as a solution of the equation (21). A view of the file containing the results of the calculus program is presented in list 1.

List 1

The results of the calculus program for flow through diaphragm

Metering system constructive data	
Pipe diameter (m)	5.0000000000E-02
Diaphragm diameter (m)	3.5000000000E-02
Diaphragm differential pressure (N/m2)	2.4525000000E+04
Fluid characteristics	
Density (kg/mc)	797.000
Viscosity (m2/s*1e-6)	3.7600000000E-06
Auxiliary parameters calculus	
Beta	7.0000000000E-01
E	1.1471541425E+00
C1	6.2390515175E-01
K	6.9005483903E+00
kod	1
Flow rate (kg/s)	4.4015625000E+00
Flow ratet (m3/s)	1.9881587202E+01

2.3.3 The algorithm for the diaphragm diameter calculus

The diaphragm calculus algorithm is derived from the calculus relation for the mass flow (10). Starting from this relation, the next nonlinear equation is drawn

$$g(d) = 0 , \tag{23}$$

where the function $g(d)$ has the expresion

$$g(d) = Q_m - C\, E \frac{\pi}{4} d^2 \sqrt{2\,\rho\,\Delta P} . \tag{24}$$

Since the factors ε, ΔP şi ρ do not depend on the diaphragm diameter d, the relation (24) can be written in the form:

$$g(d) = Q_m - K\, C\, d^2 , \tag{25}$$

where

$$K = \varepsilon\, \pi \frac{\sqrt{2\,\rho\,\Delta P}}{4} . \tag{26}$$

The diaphragm diameter calculus algorithm was transposed into a calculus program.

2.3.4 Industrial application concerning the diaphragm diameter calculus

A flow metering system is considered, having the characteristics:

- Pipe diameter = 50 mm
- Measuring domain of the differential pressure transducer = 1000 mm H_2O
- Fliud density = 940 kg/m³
- Fluid viscisity = 10.6×10^{-6} m²/s
- Fluid flow rate = 1.1 kg/s.

The request is the determination of the diaphragm diameter corresponding to the flow rate and to the known elements of the flow measuring system. The numerical results of the calculus program are presented in list 2.

List 2

The results of the diaphragm diameter calculus program

The constructive data of the metering system
Pipe diameter	(m)	5.0000000000E-02
Diphragm differential pressure (N/m2)		2.4525000000E+04

Fluid characteristics
Density	(kg/mc)	797.000
Viscosity	(m2/s*1e-6)	4.6700000000E-06
Flow rate	(kg/s)	1.5000000000E+00
Diameter	(m)	2.2250000000E-02

2.4 Flow transducers with tips

A particular case is represented by the air flow metering at the pipe furnaces. Since the pipe furnaces or the steam heaters are provided with air circuits having a rectangular section, there are no conditions for the diaphragm flow transducers to be used. For this pipe section type, there are not provided any calculus prescriptions. In this case, there has been analysed the adaptation of the long radius tip in the conditions. A cross-section through the sensitive element is presented in figure 8.

As the long radius tip is characterized by a continous and smooth variation of the throttling element, there is justified the hypothesis according to which the pressure drop for this sensitive element is due to the effective decrease of the following section.

The calculus relations for the design of the sensitive element are derived from the relation (10), written in the form

$$Q_m = a A_0 \sqrt{2 \Delta P \rho} \text{ ,[kg/s]} \tag{27}$$

where α represents the flow coefficient; A_0 – the maximum flowing area; ΔP – the drop pressure between the upstream and downstream plugs of the sensitive element.

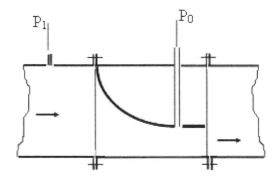

Fig. 8. The geometry of the long radius tip

By combining the relation (27) with the relation (15) there is obtained

$$Q_v = \alpha A_0 \sqrt{\frac{2\Delta P}{\rho}} \cdot [\text{m}^3/\text{s}] \tag{28}$$

The design of the sensitive element presented in figure 8 means determining the value of area A_0. In terms of the hypotheses enumerated at the beginning of the section, the calculus algorithm for the flow transducer dimensioning consist of the following calculus elements:

- Bernoulli equation

$$P_1 + \frac{\rho_1 w_1^2}{2} = P_0 + \frac{\rho_0 w_0^2}{2} \tag{29}$$

- The mass conservation equation

$$\rho_1 w_1 A_1 = \rho_0 w_0 A_0. \tag{30}$$

Since the density variation is nonsignificant for the difference in 100 mm CA, $\rho_0 = \rho_1$ is considered, that leads to

$$P_1 - P_0 = \Delta P = \frac{\rho_1 w_0^2}{2}\left(1 - \frac{w_1^2}{w_0^2}\right). \tag{31}$$

By combining the relations (28), (29), (30) and (31) there is obtained the expression used at the design of the flow transducer based on the long radius tip

$$A_0 = A_1 \sqrt{\frac{1}{1 + \frac{2\Delta P}{\rho}\left(\frac{A_1}{Q_0}\right)^2}} \cdot \tag{32}$$

The standards in the domain of the Venturi type flow transducers specify an ellipse spring for the diaphragm profile, described by the equation

$$\frac{x^2}{a^2} + \frac{y^2}{b^2} = 1 ,$$ (33)

where a and b are the demi-axis of the ellipse, figure 9.

The ellipse quotes, the pairs of coordinates points (x, y) can be calculated from the relation (33), where the variable x has discrete values in the domain $[0, a]$.

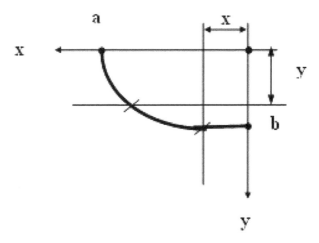

Fig. 9. Cross-section through long radius tips

2.5 Case study: The design of the long radius tip for an air flow metering system

There is considered a steam furnace within a catalytic cracking unit (CO Boyler). The initial design data is presented in table 6. The request is the dimensioning of the sensitive element of the air flow transducer.

Specification	Value
The pipetubes section profile upstream the sensitive element	1094 mm x 1094 mm
Total length of the part that contains the sensitive element	1100 mm
Maximum air flow	75000 m^3_N/h
Air pressure upstream the sensitive element	100 mm CA relativ
Maximum pressure drop on the sensitive element	100 mm CA
Air temperature	20° C

Table 6. Steam furnace design data

Solution. The problem solution has been obtained by passing through the following stages:

1.   The air flow calculus in conditions of flowing through the sensitive element.

2.   Determination of the area and of the minimum flowing limit.
3.   The calculus of the coordinates of the sensitive element component ellipse.

Stage 1. The air flow calculus in the conditions of flowing through the sensitive element:

-   air density in the conditions of flowing into the sensitive element

$$\rho_N = \frac{M}{RT_N} P_N = \frac{28.8}{8314 \times 273} \times 13590 \times 9.81 \times 0.76 = 1.286 \quad \left[kg / m_N^3 \right];$$

-   air density in the conditions of flowing into the sensitive element (P_0, T_0)

$$\rho_0 = \frac{M}{RT_0} P_0 = \frac{28.8}{8314 \times 293} \times 13590 \times 9.81 \times 0.767 = 1.209 \quad \left[kg / m^3 \right];$$

-   volumetric flow in the conditions (P_0, T_0)

$$Q_0 = \frac{\rho_N}{\rho_0} Q_N = \frac{1.286}{1.209} \times 75000 = 79777 \quad \left[m^3 / h \right];$$

$$Q_0 = \frac{79777}{3600} = 22.1602 \quad \left[m^3 / s \right].$$

Stage 2. The determination of the area and the minimum flow limit value:

-   pipetubes area calculus

$$A_1 = 1.094 \times 1.094 = 1.1968 \quad \left[m^2 \right];$$

-   calculus of pressure drop on the sensitive element

$$\Delta P = 1000 \times 9.81 \times 0.1 = 981 \quad \left[N / m^2 \right];$$

-   calculus of the minimum section A_0

$$A_0 = 1.1968 \sqrt{\frac{1}{1 + \frac{2 \times 981}{1.209} \times \left(\frac{1.1968}{22.1602} \right)^2}} = 0.4998 \quad \left[m^2 \right];$$

-   calculus of the minimum flow limit, figure 9

$$S = \frac{0.4998}{1.094} = 0.456 \quad [m].$$

Based on the result obtained, there is adopted S = 460 mm.

Stage 3. The calculus for the coordinates component ellipse of the sensitive element is based on the relation (33). From figure 10 and from the data presented in table 9 there result the following dimensions:

- the big demi-axi of the ellipse

$$a = 1100 - 300 - 50 = 750 \quad [mm] \ ;$$

- the small demi-axis of the ellipse

$$b = 1094 - 460 = 634 \quad [mm] \ .$$

By using the values of the ellipse, the relation (33) becomes

$$\frac{x^2}{0.75^2} + \frac{y^2}{0.634^2} = 1 \ .$$

In table 7 there are presented the values of the points that define the profile of the ellipse spring.

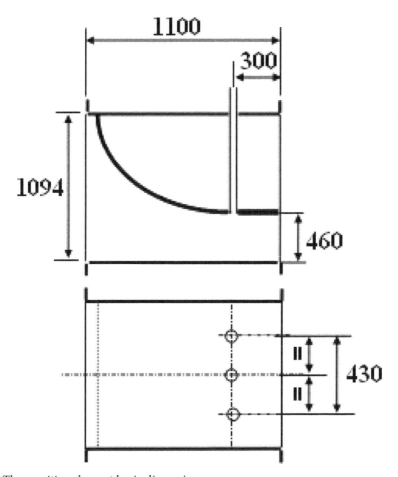

Fig. 10. The sensitive element basic dimensions

x [mm]	y [mm]	$1094-y$ [mm]	$(1094-y)$ [mm]
0	634	460	460
50	632	461	460
100	628	465	465
150	621	472	470
200	611	482	480
250	598	496	495
300	581	513	515
350	560	533	535
400	536	558	560
450	507	587	590
500	472	621	620
550	431	663	665
600	380	714	715
650	316	778	780
700	227	866	865
750	0	1094	1094

Table 7. The values of the ellipse spring profile

3. Fluid flow control systems based on control valves

3.1 The structure of the control valves

The control valves are the most widely-spread control elements within the chemical, oil industry etc. For these cases, the execution element is considered a monovariable system, the input quantity being the command u of the controller, the output quantity being identified with the execution quantity m, associated to the process. Taking into consideration the fact that the command signal u is an electrical signal in the range $[4...20]$ mA, the drive of the control element needs a signal convertor provided with a power amplifier, a servomotor and a control organ specifical to the process. In figure 11 there is presented the structure of an actuator of control valve – type. This is made of an electrical – pneumatic convertor, a pneumatic servomotor with a membrane and a control valve body with one chair.

The electro – pneumatic convertor changes the electrical signal u, an information bearer, into a power pneumatic signal p_c. The typical structure of this subsystem is presented in figure 12 (Marinoiu et. al. 1999). This contains an electromagnet (1), a permanent magnet (3), a pressure – displacement sensor (2), a power amplifier (4) and a reaction pneumatic system (5). A lever system ensures the transmission of information and of the negative reaction into the system.

The other two subsystems, the servomotor and the control valve body, are interdependent, their connection being of both physical and informational nature. In the drawing of the figure 13 there is presented a servomotor with a diaphragm, a servomotor that ensures a normally closed state of the control valve. The contact element between the two subsystems is represented by the rod (3). This transmits the servomotor movement, expressed by the rod h displacement, towards the control valve body. The command pressure of the servomotor, p_C, represents the output variable of the electro – pneumatic convertor. The control valve body represents the most complex subsystem within the control valve. This will modify the servomotor race h and, accordingly, the valve plug position in ratio with the control chair. The change of the section and the change of the flowing conditions in the control valve body will lead to the corresponding change of the flow rate.

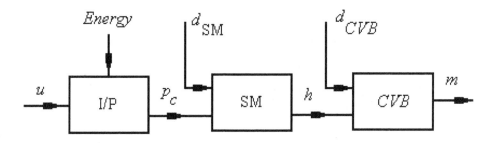

Fig. 11. The component elements of a control valve: E/P – electrical – pneumatic convertor; SM – servomotor; CVB – control valve body; u – electrical command signal; d_c – pneumatic command signal; h – servomotor valve travel; d_{SM} – disturbances associated to the servomotor; d_{CVB} – disturbances associated to the control valve body.

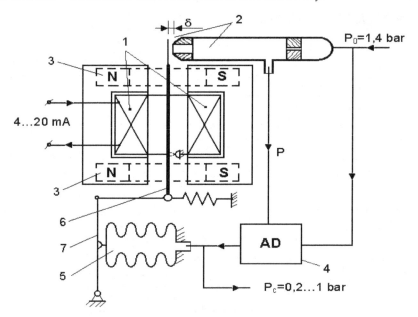

Fig. 12. The electro – pneumatic convertor: 1 – electromagnetic circuit; 2 – the pressure – displacement sensor; 3 – permanent magnets; 4 – power amplifier; 5 – the reaction bellows; 6 – articulate fitting; 7 - lever.

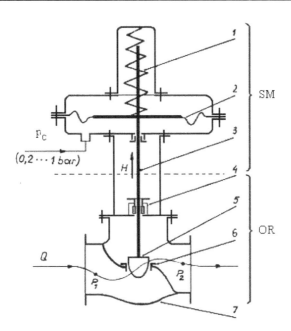

Fig. 13. The servomotor – control organ subsystem: SM – servomotor; CVB- control valve body; 1 - resort; 2 – rigid diaphragm; 3 - rod; 4 - sealing system; 5 - valve plug stem; 6 – chair; 7 - body.

3.2 The constructive control valve types

The usually classification criteria of the control valves are the following (Control Valve Handbook, Marinoiu et al. 1999):

a. The valve plug system:
   - a profiled valve;
   - a profiled skirt valve or a valve with multiple holes;
   - a cage with V-windows;
   - a cage with multiple holes;
   - special valve plug systems;
   - no valve plug systems;
b. The ways of the fluid circulation through the control organ:
   - straight circulation;
   - circulation at 90° (corner valves);
   - divided circulation (valves with three ways).
c. Numbers of chairs:
   - a chair;
   - two chairs.
d. The constructive solution imposed by the nature, temperature and flowing conditions:
   - normal;
   - with a cooling lid with gills;

- with a sealing bladder;
- with an intermediate tube;
- with a heating mantle.

3.3 The control valves modelling

The modeling of the control valves represents a delicate problem because of the complexity design of the control valves, because of the hydraulic phenomena and the dependency between the elements of the control system: the process, the transducer, the controller and the control valve.

From the hydraulic point of view, the control valve represents an example of hydraulic variable resistor, caused by the change of the passing section. An overview of a control valve, together with the main associated values, is presented in figure 14. When the h movement of the valve plug modifies, there results a variation of the drop pressure ΔP_v and of the flow Q which passes through the valve.

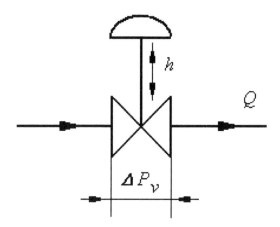

Fig. 14. Overview of a control valve: h – the movement of the valve plug's strangulation system; Q – the debit of the fluid; ΔP_v - the drop pressure on the control valve.

3.3.1 The inherent valve characteristic of the control valve body

The inherent valve characteristic of the control valve body represents a mathematical model of the control valve body that allows the determination, in standard conditions, of some inherent hydraulic characteristics of the control valve, irrespective of the hydraulic system where it will be assembled. A control valve can not always assure the same value of the flow Q for the same value of movement h, unless there is an invariable hydraulic system. This aspect is not convenient for modeling the control valve as an automation element, because it implies a different valve for every hydraulic system. A solution like this is not acceptable for the constructor, who should make a control valve for every given hydraulic system The inherent characteristic represents the dependency between the flow modulus of the control valve body and the control valve travel

$$K_v = f(h) \tag{34}$$

The flow modulus K_v represents a value that was especially introduced for the hydraulic characterization of the control valves, its expression being

$$K_v = \sqrt{2}\alpha A_v \quad [\text{m}^2], \tag{35}$$

where A_v – the flow section area of the control valve; α – the flow coefficient.

The way K_v value was introduced through relation (35) shows that it depends only on the inherent characteristics of the control valve body, which are expressed based on its opening, so based on the movement h of the valve plug. Keeping constant the drop pressure on the valve, there is eliminated the influence of the pipe over the flow through the control valve and the dependency between the flow and the valve travel is based only on the inherent valve geometry of the valve.

The inherent valve characteristics depend on the geometric construction of the valve control body. Geometrically, the valve control body can be: a valve plug with one chair, a valve plug with two chairs, a valve plug with three ways, a valve plug especially for corner valve etc. Consequently, the mathematical models of the inherent valve characteristics will be specific to every type of valve plug.

In the following part, there will be exemplified the mathematical models of the inherent valve characteristics for the valve control body with a plug valve with one chair. For this type of valve plug, there are used two mathematical models, named *linear characteristic* and *logarithmical characteristic*, models which are defined through the following relations:

- linear characteristic dependency

$$\frac{K_v}{K_{vs}} = \frac{K_{v0}}{K_{vs}} + \left(1 - \frac{K_{v0}}{K_{vs}}\right)\frac{h}{h_{100}} ; \tag{36}$$

- logarithmical characteristic dependency

$$\frac{K_v}{K_{vs}} = \frac{K_{v0}}{K_{vs}}\exp\left(\frac{h}{h_{100}}\ln\frac{K_{vs}}{K_{v0}}\right), \tag{37}$$

where h is the movement of the valve plug related with the chair; h_{100} – the maximum value of the plug's valve travel; K_{v0} – the value of K_v for $h = 0$; K_{vs} – the value of K_v at maximum valve travel h_{100}.

In figure 15 there are presented the graphical dependency for the two mathematical models of the inherent characteristics of the valve control body with a plug valve with one chair (Marinoiu et. al. 1999).

Observations. The value K_v, used within the mathematical model of the inherent valve characteristic and for the hydraulic measurement of the control valves, was introduced by Früh in 1957 (Marinoiu et al. 1999). Through the relation (35) he shows that the flow modulus K_v has an area dimension; out of practical reasons there has been agreed to be attributed to K_v a physical meaning, which would lead to a more efficient functioning. This new meaning is based on the relation

$$K_v = \frac{Q}{\sqrt{\dfrac{\Delta P_r}{\rho}}} \quad [\text{m}^2], \tag{38}$$

and has the following interpretation:

K_v is numerically equal with a fluid of $\rho = 1$ kg/dm³ density which passes through the control valve when there takes place a pressure drop on it of $\Delta P_r = 1$ bar. The numerical values of K_v are expressed in m³/h .

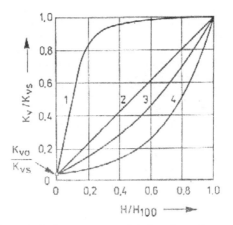

Fig. 15. Inherent valve characteristics types associated to the valve control body with a valve plug with one chair: 1 – fast opening; 2 – linear characteristic; 3 – equally modified percentage; 4 – logarithmical characteristic.

In the USA, by replacing the value K_v there is defined the value C_v as being the water flow expressed in gallon/min, which passing through the control valve produces a pressure drop of 1 psi. The transformation relations are the following:

$$\begin{cases} C_v = 1,156K_v \\ K_v = 0,865C_v \end{cases} ; \tag{39}$$

3.3.2 The work characteristic of the control valve body

The work characteristic of the control valve represents the dependency between the flow Q and the valve travel of the h valve plug

$$Q = Q(h). \tag{40}$$

When defining the static work characteristic there is no longer available the restrictive condition concerning the constant pressure drop on the valve, as it was necessary for the inherent valve characteristic, but the flow rate gets values based on the hydraulic system where it is placed, the size, the type and the opening of the valve control. From the point of view of the hydraulic system, the working characteristics can be associated to the following systems:

a.   systems without branches;
b.   hydraulic systems with branches;
c.   hydraulic systems with three ways valves.

Due to the phenomena complexity, for the mathematical modeling of the working characteristic of the valve control body, there are introduced the following simplifying hypotheses:

a.   there is taken into consideration only the case of the indispensable fluids in turbulent flowing behavior ;
b.   there are modelled only the hydraulic systems without branches;
c.   the loss of pressure on the pipe is considered a concentrated value.

The main scheme of a hydraulic system without ramifications is presented in figure 16. The system is characterized by the loss of pressure on the control valve ΔP_v, the loss of pressure on the pipe ΔP_p and the loss of pressure inside the source of pressure ΔP_{SI}.

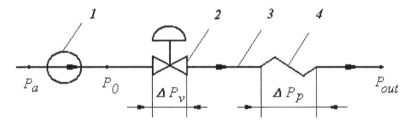

Fig. 16. Hydraulic system without branches: 1 - pump; 2 – control valve; 3 - pipe; 4 – the hydraulic resistance of the pipe.

For the modeling, the working characteristic of the valve control body, are defined by the following values:

•   The flow rate that passes through the valve control

$$Q = K_V \sqrt{\frac{\Delta P_v}{\rho}} \quad [\text{m}^3/\text{h}]. \tag{41}$$

•   The energy balance of the hydraulic system

$$P_0 = P_{out} + \Delta P_v + \Delta P_p. \tag{42}$$

The connection of the control valve with the hydraulic system is very tight. To be able to determine the working characteristics of the control valve there have to be solved all the elements of modeling presented in this chapter: the centrifugal pipe characteristic, the inherent valve characteristic of the control valve and the pipe characteristic. The mathematical model of the control valve working characteristic is defined by the block scheme presented in figure 17. The input variable is the valve travel h of the servomotor and implicit of the control valve and the value of exit is the flow rate Q which passes through the valve. Mathematically, the model of the control valve presented in figure 17 is a nonlinear equation

$$f(Q) = Q - K_v \sqrt{\frac{\Delta P_v}{\rho}} = 0 . \qquad (43)$$

3.3.3 Solutions associated to the working characteristic of the control valve

The working characteristic of the control valve body, materialized by relation (40), can be determined in two ways:

a. By introducing the simplifying hypothesis according to which there is considered that the flow modulus associated to the pipe does not modify, respectively ;
b. By resolving numerically the model presented in figure 17.

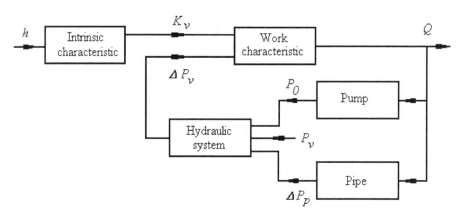

Fig. 17. The block scheme of the mathematical model of the body control valve.

The solution obtained by using the simplifying hypothesis represents the classical mode to solve the working characteristics of the control valve body (Früh, 2004). The solution has the form (Marinoiu et. al. 1999)

$$q = \frac{1}{\sqrt{1 + \Psi\left(\frac{1}{k_v^2} - 1\right)}} , \qquad (44)$$

where $q = Q/Q_{100}$ represents the adimensional flow rate; $k_v = K_v/K_{v100}$ - the adimensional flow module; $\Psi = \Delta P_{v100}/\Delta P_{hs}$ - the ratio between the maximal control valve drop pressure and the maximal hydraulic system drop pressure.

In figure 18 there are presented the graphical solution of the working characteristics of the control valves for the valve plug type with inherent valve linear characteristic and inherent valve logarithmic characteristics.

The solution obtained by the numerical solving of the mathematical model presented in figure 17 has been recently obtained (Patrascioiu et al. 2009). Unfortunately, the mathematical model and the software program are totally dependent on the centrifugal pump, pipe and the control valve type (Patrascioiu 2005). In the following part there are presented an example of the hydraulic system model and the numerical solution obtained. The hydraulic system contains a centrifugal pump, a pipe and a control valve. The pump characteristic has been presented in figure 3 and the mathematical model of the 50-20 pump type is presented in table 1. The pipe of the hydraulic system has been presented in table 2 and the pipe mathematical model is expressed by the relation (2). The control valve of the hydraulic system is made by the Pre-Vent Company, figure 19, the characteristics being presented in table 8 (www.pre-vent.com).

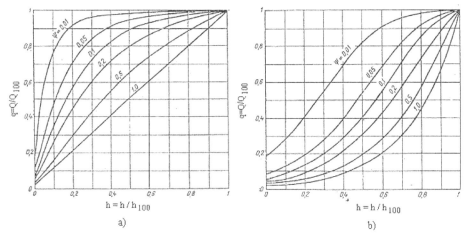

Fig. 18. The working characteristics of the control valves calculated (based on the simplifying hypothesis): a) valve plug with linear characteristic; b) valve plug with logarithmic characteristic.

The numerical results of the program are the inherent valve and the work characteristics. For theses characteristics, the independent variable is the adimensional valve travel of the control valve, $h/h_{100} \in [1...100]\%$.

Variable	Measure unit	Value
Inherent valve characteristic	Linear	
K_{vs}	m³ h⁻¹	25
K_{v0}	m³ h⁻¹	1

Table 8. The Control valve characteristics.

The inherent valve characteristic obtained by calculus confirms that the control valve belongs to the linear valve plug type. The working characteristic of the control valve is almost linear, figure 20. The pipe drop pressure is very small, figure 5, and for this reason the influence of the control valve into hydraulic system will be very high, see figure 21. In this context, the inherent valve characteristic of the control valve body is approximately linear.

Fig. 19. The control valve made by Pre-Vent Company.

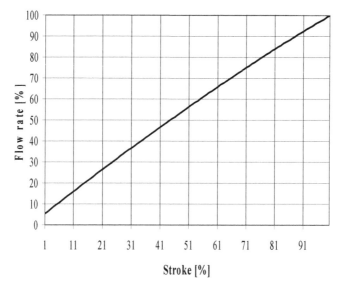

Fig. 20. The work characteristic of the control valve from the studied hydraulic system.

The picture presented in figure 21 is similar to the output pressure of the pump. The conclusion resulting is that 99% of the hydraulic pump energy is lost into the control valve. For this reason, the choice of a control valve with linear characteristicis wrong, the energy being taken into consideration.

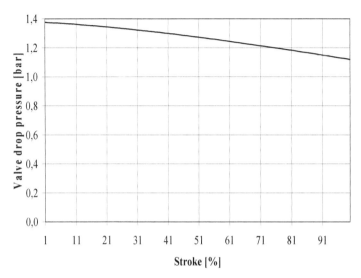

Fig. 21. The control valve drop pressure.

3.4 The control valve design and the selection criteria

The control valves are produced in series, in order to obtain a low price. For this goal, the control valve producers have realized the proper standards of the geometric and hydraulic properties. The control valves choice is a complex activity, composed of technical, financial and commercial elements. Mainly, the control valves choice represents the selection of a type or a subtype industrial data based on a control valve, depending of one or many selection criteria.

Technical criteria refer to the calculus of the technical parameters of the control valves. The financial elements include the investment value and the operation costs. The commercial elements describe the producers 'offers of various types of control valves.

The technical parameters of the control valves contain at least the flow module calculus of the control valve of the control system. The choice of the control valve involves the following elements: the constructive type of the control valve body, the standard flow module K_{vs} of the control valve manufactured by a control valve company and the nominal diameter D_n of the control valve.

3.4.1 The flow module calculus

The design relations of the control valves are divided in two categories: classical relations and modern relations. The classical relations have been introduced by Früh (Früh 2004). Theses relations are recommended for the design calculus of the control valves placed in the hydraulic systems characterized by turbulent flow regime, hydraulic system characterized by without branches and for the calculus initialization of the other design algorithms. A short presentation of these relations is presented in table 9.

ΔP_v	Fluid type		
	Liquid	Gas	Overheated steam
$\Delta P_v < \dfrac{P_1}{2}$	$K_v = Q\sqrt{\dfrac{\rho}{\Delta P_v}}$	$K_v = \dfrac{Q_N}{514}\sqrt{\dfrac{\rho_N T_1}{P_2 \Delta P_v}}$	$K_v = \dfrac{Q_m}{31.6}\sqrt{\dfrac{v_2}{\Delta P_v}}$
$\Delta P_v > \dfrac{P_1}{2}$		$K_v = \dfrac{Q_N}{257}\sqrt{\dfrac{\rho_N T_1}{P_1}}$	$K_v = \dfrac{Q_m}{31.6}\sqrt{\dfrac{2v_2}{P_1}}$
Notation and measure units	Q – volumetric flow rate [m³/h]	Q_N – volumetric flow rate [m³$_N$/h]	Q_m – mass flow rate [kg/h]
	-	P_1 – upstream pressure [bar]; P_2 – downstream pressure [bar]	P_1 – upstream pressure [bar]
	ΔP_v - control valve drop pressure [bar]		
	ρ - density [kg/dm³]	ρ_N – normal density [kg/m³$_N$]	v_2 – specific volume of the steam [m³/kg]

Table 9. The classical design relations for control valve flow module.

The modern relations of the flow module are based on the ISA standards and are characterized by all flow regimes (laminar regime, crossing regime, turbulent regime, cavitational flow regime) and for a more hydraulic variety (ISA 1972, 1973). Also, the calculus relations are specific to the fluid type (incompressible fluid and compressible fluid).

3.4.2 The industrial control valves production

The control valves companies produces various types of standardized control valves. Each company has the proper types of control valves and each control valve type is produced a various but standardized category, defined by standardized flow module, nominal diameter and chair diameter. Each company presents their control valves offer for chemical and control engineering. In figure 22 there is presented an image of the BR-11 control valve type from the Pre-Vent Company. There are presented the standard flow module, the maximal valve travel and the offer of nominal diameter of the control valve.

3.4.3 The control valves choice criteria applied to the flow control system

The control valves choice represents an important problem of the control systems design. The control valves choice criteria are the following:

a. For each control system, there must be chosen an inherent valve characteristic of the control valve body (or a control valve type) so that all the components of the control system generate the lowest variation of the control system gain;
b. For each control system there must be chosen a working characteristic of the control valve body so that all the components of the control system to generate a linear characteristic.

The common flow control system has the structure presented in figure 1. Using the previous choice criterion, some recommendations can be made for the selection of the control valve of the flow control systems, table 10 (Marinoiu 1999). The flow control system can meet the stabilization control function or the tracking control function.

Kv_s (m³/h)	Stroke (mm)	Plug face (mm)	Valve nominal size DN											
			15	20	25	32	40	50	65	80	100	150	200	250
0,010														
0,016														
0,025														
0,040														
0,063														
0,10		6,35												
0,16														
0,25														
0,40														
0,63	20													
1,0														
1,6		9,52												
2,5		12,70												
4,0														
6.3		19,05												
10		20,64												
16		25,25												
25		31,72												
40		41,25												

Fig. 22. The BR-11 control valve type offer of the Pre-Vent Company

Control system type	Flow transducer characteristic	Ψ	Disturbances	Inherent recommended characteristic
stabilization	$r \approx Q$	1	P_{r0}, P_v, T, v, ρ	logarithmic
	$r \approx Q^2$			linear only if $Q_{i1} \neq Q_{i2}$
	$r \approx Q$	<1	$P_{r0}, P_v, R_1, R_2, T, v, \rho$	logarithmic
	$r \approx Q^2$			
tracking	$r \approx Q$	1	The disturbance variations are negligible as compared to the set point variations	linear
	Linearized by square root	≤0,3		linear
		≤0,3		logarithmic
	$r \approx Q^2$	1		logarithmic
	Non linearized	<1		linear

Table 10. Recommendations for the choice of the control valves for the flow control systems.

4. References

Control Valve Handbook, Fourth Edition, Emerson Process Management,
    www.documentation.emersonprocess.com/groups/public/
Früh, K. F. (2004). *Handbuch der Prozessautomatisierung*, R. Oldenbourg Verlag, München.
http://www.pre-vent.com/en/br11.html
ISA-S 39.1 (1972), *Control valve Sizing Equations for Incompressible Fluids*.
ISA 39.3 (1973), *Control valve Sizing Equations for Compressible Fluids*.
Marinoiu V. Poschina I., Stoica M., Costoae N. (1999), *Control elements. Control valves*,
    (Edition 3), Editura Tehnica, ISBN 973-31-1344-1, Bucuresti, (Romanian).
Pătrăşcioiu C. (2005). *Numerical Methods applied in Chemical Engineering – PASCAL
    Applications*, (Edition 2), Editura MatrixRom, ISBN 973-685-692-5, Bucuresti,
    (Romanian).
Patrascioiu, C., Panaitescu C. & Paraschiv N. (2009). Control valves – Modeling and
    Simulation, *Procedings of the 5th WSEAS International Conference on Dynamical
    Systems and Control (CONTROL 09)*, pp. 63-68, ISBN 978-960-474-094-9, ISSN 1790-
    2769, LaLaguna, Spain, 2009.

5

Strategies to Increase Energy Efficiency of Centrifugal Pumps

Trinath Sahoo

M/S Indian Oil Corporation ltd., Mathura Refinery, India

1. Introduction

As the pie chart below ndicates, the LCC of a typical industrial pump over a 20 year period is primarily made up of maintenance and energy costs.

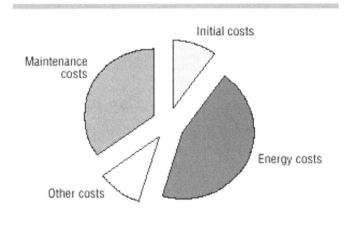

Fig. 1. Life cycle cost of centrifugal pump

Energy cost is highest of the total cost of owning a pump. Centrifugal pumps consume, depending on the industry, between 25 and 60% of plant electrical motor energy.

INDUSTRY	PUMP ENERGY (% of Total Motor Energy)
Petroleum	59%
Pulp & Paper	31%
Chemical	26%

Proper matching of pump performance and system requirements, however, can reduce pump energy costs by an average of 20 percent in many cases. The process of specifying the right pump technology for an application in facilities should go well beyond first cost, but in too many cases, it does not. Such a shortsighted approach can create major, long-term problems for organizations.

In the process industries, the purchase price of a centrifugal pump is often 5 - 10% of the total cost of ownership. Typically, considering current design practice, the life cycle cost (LCC) of a 100 horsepower pump system, including costs to install, operate, maintain and decommission, will be more than 20 times the initial purchase price. In a marketplace that is relentless on cost, optimizing pump efficiency is an increasingly important consideration.

Once the pump is installed, its efficiency is determined predominately by process conditions. The major factors affecting performance includes efficiency of the pump and system components, overall system design, efficient pump control and appropriate maintenance cycles. To achieve the efficiencies available from mechanical design, pump manufacturers must work closely with end-users and design engineers to consider all of these factors when specifying pumps.

Analysis of different losses encountered in pump operation:-

1.   Mechanical friction loss between fixed and rotating parts:-

The main components are:-

- The external sleeve, ball or roller journal bearings.
- The internal sleeve bearings.
- Thrust bearings.
- The gland, neck bush and packing rings.

 The proportion of total mechanical loss contributed by each of these will depend upon

the type and condition of the pump. Proper lubrication of the bearings and glands shall reduce the frictional losses.

2.   Disc friction loss between the liquid and the external rotating faces of the rotor discs:-

In principle the parts concerned here are all those rotating surfaces of the pump in contact with the liquid that do not actually take share in guiding the liquid. Such elements are:-

- The external faces of the rotor discs or shroud.
- The outer edges of the shroud.
- The edges of the sealing rings.
- The whole surface of balanced discs.

By measuring the power needed to drive the disc in a variety of conditions it appears that this power loss depends upon rotational speed, disc diameter, roughness of the sides of the discs and inner walls of the casing, the density and viscosity of the liquid and axial clearance between the disc and casing.

The axial clearance affect the power loss. Any element of liquid in contact with revolving disk will be dragged round with it, at least for a short distance and during this journey the element will necessarily be subjected to centrifugal force. This will induce it to slide outwards. Other elements from the main body of the liquid will flow into replace the original one and hence additional kind of flow will be set up as shown in **fig. below.**

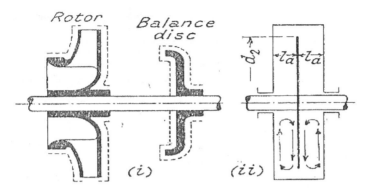

Fig. 2. Diagaram showing areas subject to disc friction loss, etc.

That is the frictional impulsion has created on a very small scale the pumping effect that the direct thrust of the impeller blades creates on an effective scale.

But it seems likely that in the space between disc and casing, the axial distance l_a will influence both the radial and tangential velocity components. If this distance is large it will be easy for relatively large amounts of liquid to become involved in the secondary circulation and thereby to steal energy from the disc. But if the distance is small, the energy should be less.

The above study shows that in comparable conditions and increase in the axial clearance causes an increase in the power loss.

Hence it can be minimized by having as small axial clearance as possible so as to improve efficiency

3.   The leakage power loss:-

The leakage liquid is bled off the main stream at a number of points, each at different pressures. Leakage may occur at other region than sealing rings:-

- Through lantern rings of liquid sealed stuffing boxes.
- Past the balance disc of multistage pumps.
- Past the neck bushes in the diaphragm of multi-stage pumps.
- Past the main glands/mechanical seals to waste.

In normal pumps only the leakage past the balance discs of multistage pumps can be measured and the pressure head at the leak off points is fairly well known. The leakage thru the glands / mechanical seals can be reduced by attending these leakage from time to time.

4.   Hydraulic power loss:-

if we have calculated the values of mechanical loss P_b, the disc friction loss P_d and leakage loss P_l, then we can assert that the residual energy P_s- P_b –P_d –P_l must wholly be transferred to the main stream of liquid flowing through the pump. There is no where else for it to go. But this is a very different thing from saying that the liquid receives the corresponding net energy increment during its passage from suction flange to delivery flange. The difference between the two quantities is what we call hydraulic power loss(P_h), it can be computed in this way.

Energy En is transferred to the liquid is utilized in imparting tangential acceleration to the liquid elements.

Unit weight of liquid will receive (V_nV_2/g) units of energy as there are W units of liquid effectively flowing per second it follows that

$$(W / K_p) * (V_nV_2)/g = P_s - P_b - P_d - P_l = P_w + P_h$$

where V_n is actual whirl component, V_2 is tangential velocity of liquid and k_p represents energy per second corresponding to one horse-power.Now the net energy increment per unit weight of liquid is represented by He and the net power received, W.H.P or P_w is WH_e/K_p.

Therefore hydraulic power loss Ph can be written

$$P_h = (W/K_p)*\{(V_nV_2/g) - H_e\}$$

And the hydraulic efficiency

$$\eta = [H_e/(V_nV_2/g)] = P_w/(P_w + P_h)$$

Losses under reduced and increased flow condition

An examination of inlet velocity triangle shows that additional energy losses are now to be reckoned with in increase flow or reduced flow conditions.

a.   In Rotor :

If the blade tips were made tangential to the designed inlet relative velocity vr1, they cannot be at the same time be tangential to the modified velocities vr1c or vr1d. Because of this eddies may form.

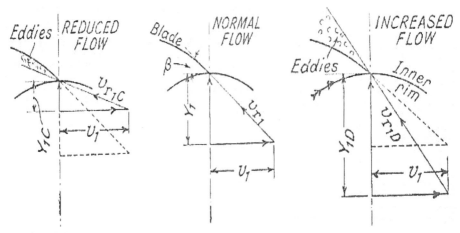

Normal and distorted inlet velocity diagrams.

When the state of flow is below normal, there may be a tendency for the liquid stream in each of impeller passages to concentrate near the front of the blades, leaving a more or less dead space near the back of the blades, that in regard to effective forward motion but alike and its capacity to waste energy.

b.   In the Recuperator :-

When the pump discharge is reduced there will be additional energy losses. While the velocity of the liquid leaving the impeller and entering the volute is above normal yet the mean velocity in the volute itself must be below normal. Thus the essential condition for maximum efficiency of energy conversion can no longer apply.

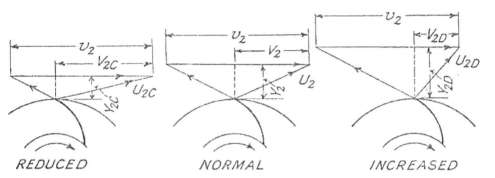

Ideal outlet velocity diagrams for reduced, normal, and increase flow in centrifugal pump impeller.

If now the pump discharge is considerably above normal, then the absolute velocity vector (figure 1) takes an abnormal inclination. Very serious contraction of main stream of liquid may occur at volute tongue (figure 2) with consequent energy losses.

In regard to guide blade or diffuser type recuperator(figure 3) shows that abnormally higher rates of flow will again destroy the necessary correspondence between vane inclination and velocity vector inclination. Energy dissipation is augmented here also

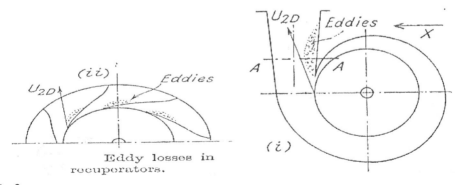

Eddy losses in recuperators.

Fig. 3.

2. Operation at high flows

The vast majority of pumping systems run far from their best efficiency point (BEP). For reasons ranging from shortsighted or overly conservative design, specification and procurement to decades of incremental changes in operating conditions, most pumps, pipes and control valves are too large or too small. In anticipation of future load growth, the end-user, supplier and design engineers routinely add 10 to 50% "safety margins" to ensure the pump and motor can accommodate anticipated capacity increases.

Under this circumstances the head capacity curve intersects the system head curve at a capacity much in excess of the required flow using excess power. Of course the pump can be throttled back to the required capacity and the power is reduced some what. But if the pump runs uncontrolled it will always run at excess flow. unless sufficient NPSH has been made available, the pump may suffer cavitation damage and power consumption will be excessive.Important energy savings can be made if at the time of selecting the condition of service, reasonable restraints are exercised to avoid using excessive safety margins for obtaining the rated service condition. But in an existing installation if the pumps have excessive margins the following options are available.

a. The existing impeller can be cut down to meet the condition of service required for the installation.
b. A replacement impeller with the necessary reduced diameter may be ordered from the pump manufacturer.
c. In certain cases there may be two separate impellers designs available for the same pump, one of which is of narrower width than the one originally furnished. A narrower replacement will have its best efficiency at at a lower capacity than the normal width impeller.

2.1 Effect of speed and its control

Speed of the centrifugal pump has marked effect on the varios losses taking place in the pump and hence it is also one the important factor which can be controlled tyo improve efficiency.

2.2 Effect of speed on various losses

Mechanical losses: In the pump the correlation between speed N and mechanical power loss P_b will be

$$P_b = K_1 . N^m$$

The value of m is between 1 & 2depending upon the type of bearing and stuffing box.

The input power varies as the cube of speed.

$$P_s = K_2 . N^3$$

So the relative mechanical loss $= P_b/P_s = K_1 . N^m / K_2 . N^3$

Disc friction loss : For a given pump using liquid the relation between speed and power loss can be given by

$$P_d = K_3. N^{2.85}$$

The relative disc friction loss $= P_d/ P_s = K_3. N^{2.85}// K_2. N^3$

It show that the value relative disc friction loss increases as the speed decreases.

Leakage: The relative leakage loss appears to be independent of the pump speed.The leakage takes place. It seems likely that the relative leakage power loss will slightly increase as the pump speed falls

Hydraulic Power loss: As the liquid flows at a rate Q through the pump passage, it undergoes a energyu loss consist of friction loss and eddy losses.

The hydraulic power loss is denoted by

$$P_h = K_{11}. N^{2.95}$$

Comparing this with the input power

$$\Delta n = K_{11}. N^{2.95}/ K_2. N^3$$

Hence the value of the relative power loss rises as the pump speed falls.

These kind of losses can be minimized by having a proper speed control mechanism.

Speed control is an option that can keep pumps operating efficiently over a broad range of flows. In centrifugal pumps, speed is linearly related to flow but has a cube relationship with power. For example, slowing a pump from 1800 to 1200 rpm results in a 33% decreased flow and a 70% decrease in power. This also places less stress on the system.

There are two types of speed control in pumps: multi-speed motors and variable speed drives. Multi-speed motors have discrete speeds (e.g., high, medium, and low). Variable speed drives provide speed control over a continuous range. The most common type is the variable frequency drive (VFD), which adjusts the frequency of the electric power supplied to the motor. VFDs are widely used due to their ability to adjust pump speed automatically to meet system requirements. For systems in which the static head represents a large portion of the total head, however, a VFD may be unable to meet system needs.

3. Variable speed drives

Pump over-sizing causes the pump to operate to the far left of its best efficiency point (BEP) on the pump head -capacity curve. Variable speed drives, assuming a low static head system, allow the pump to operate near its best efficiency point (BEP) at any head or flow. In addition, the drive can be programmed to protect the pump from mechanical damage when away from BEP -- thereby enhancing mechanical reliability. Furthermore, excessive valve throttling is expensive and not only contributes to higher energy and maintenance cost, but can also significantly impair control loop performance. Employing a throttled control valve, less than 50% open, on the pump discharge may accelerate component wear, thereby slowing valve response.

VFDs allow pumps to run at slower speeds with further contributions to pump reliability and significant improvement in mean-time-between-failure (MTBF).

4. Effect of specific speed

The higher the specific speed selected for a given set of operating condition, the higher the pump efficiency and therefore the lower the power consumption. Barring other considerations the tendency should be to favour higher specific speed selection from the point of view of energy conservation

4.1 Clearance

Good wear ring with proper clearance improves pump reliability and reduce energy consumption. Correct impeller to volute or back plate clearance is also to be maintained. Pump efficiency decreases with time because of wear. A well designed pump usually comes with a diametral clearance of 0.2 to 0.4%. however as long as it remains below 0.6 to 0.8 % its effect on efficiency remains negligible. When the clearance start to increase beyond these values efficiency start to drop drastically. For equal operating condition the rate of wear depends primarily on design and material of the wear ring. Generally for non corrosive liquids, the resistance to wear increases with hardness of the sealing surface material.

If a pump having a specific speed of 2500, the leakage loss in a new pump will be about 1%. Thus when the internal clearances will have increased to the point that this leakage will have doubled, we can regain approximately 1% in power saving by restoring the pump clearance. But if we are dealing with a pump having a specific speed of 750, it will have leakage loss of about 5%. If the clearances are restored after the pump has worn to the point that its leakage losses have doubled, we can count on 5% power saving.

4.2 Change in surface roughness

Depending on the material of construction and properties of the liquid being pumped, the roughness of the flow path can also change over time. In some instances, the channels may acquire a smooth polish, and in other they may become roughened. Both of these changes can significantly affect pump performance. An increase in casing roughness usually reduces both the total head and the efficiency.

4.3 Change in flow path size

The dimension of the pump's flow path may change over time due to abrasion or erosion, which usually increases the pathways' dimension or to scale, rust or sedimentation which usually reduces the size of the pathways. The latter is particularly apt to occur in pumps operating intermittently.

4.4 Run one pump instead of two

Many installations are provided with two pumps operating in parallel to deliver the required flow under full load. Too often, both pumps are kept on line even when demand drops to a point where a single pump can carry the load. The amount of energy wasted in running two pumps at half load when a single pump can meet this condition is significant.

If we want to reduce the flow to half load and still maintain both pumps online, it will be necessary to throttle the pump discharge and create a new system head curve. Under these condition, each pump will deliver 50% of the rated capacity at 117% rated head, much of which will have to be throttled. Each pump will take 72.5% of its rated power consumption. Thus the total power consumption of two pumps of two pumps operating under half load condition would be 145% of that required if a single pump were to be kept on line.

5. Viscosity of liquid

Liquid viscosity affect pump performance. This is because two of the major losses in a centrifugal pump are caused by fluid friction and disc friction. These losses vary with the viscosity of the liquid being pumped, so that both the head capacity output and the mechanical output differ from the original values. As viscosity of the liquid increases the head developed by the pump decreases and the efficiency decreases. So in process industries it is required to maintain good insulation and steam tracing in the suction line of the pump.

5.1 Effect of cavitation

Due to cavitation the vapor bubbles are formed and these bubbles will modify the velocity distribution as well as the pressure distribution in the rotor passage. The effect of vapor pressure bubbles will virtually lower the density of the liquid; that is the pump behaves as if another liquid of less density were flowing through it. As the mean velocity of the fluid mixture is now at higher (for the weight per second is unchanged) the outlet velocity diagram tends to assume the distorted shape and shall lower the total head. The fall in density reduces the pressure generated.

Due to cavitation there will be loss of energy.

5.2 Effect of change of density

The pump develops the same head in meters of liquid independent of specific gravity. The pump delivers the same quantity by volume independent of specific gravity but the quantity by weight will be proportional to specific gravity. The input power and the output power for a given volumetric discharge vary in proportion to density.

If the water is heated the density changes. In high pressure steam boilers, the reduction in density of the water may be 15 percent or more. If a given weight per second of water is to be discharged against a stipulated pressure, the power output will increase as the water temperature increases.

5.2.1 Pump sizing

Selecting a centrifugal pump can be challenging because these pumps generate different amounts of flow at different pressures. Each centrifugal pump has a "best efficiency point" (BEP). Ideally, under normal operating conditions, the required flow rate will coincide with the pump's BEP.

The complexity associated with selecting a pump often results in a pump that is improperly sized for its application. Selecting a pump that is either too large or too small can reduce system performance. Undersizing a pump may result in inadequate flow, failing to meet system requirements. An oversized pump, while providing sufficient flow, can produce other negative consequences; higher purchase costs for the pump and motor assembly; higher energy costs, because oversized pumps operate less efficiently; and higher maintenance requirements, because as pumps operate further from their BEP they experience greater stress; ironically, many oversized pumps are purchased with the intent of increasing system reliability.

Unfortunately, conservative practices often prioritize initial performance over system life cycle costs. As a result, larger-than-necessary pumps are specified, resulting in systems that do not operate optimally. Increased awareness of the costs of specifying oversized pumps should discourage this tendency.

5.2.2 Variable loads

In systems with highly variable loads, pumps that are sized to handle the largest loads may be oversized for normal operating loads. In these cases, the use of multiple pumps, multi-speed motors, or variable speed drives often improves system performance over the range of operating conditions.

To handle wide variations in flow, multiple pumps are often used in a parallel configuration. This arrangement allows pumps to be energized and de-energized to meet system needs. One way to arrange pumps in parallel is to use two or more pumps of the same type. Alternatively, pumps with different flow rates can be installed in parallel and configured such that the small pump—often referred to as the "pony pump"—operates during normal conditions while the larger pump operates during periods of high demand.

5.3 Valves and fittings

Pumping system controls should be evaluated to determine the most economical control method. High-head-loss valves, such as globe valves, are commonly used for control purposes. Significant losses occur with these types of valves, however, even when they are fully open. If the evaluation shows that a control valve is needed, choose the type that minimizes pressure drop across the valve. Pumping system control valve inefficiencies in plant processes offer opportunities for energy savings and reduced maintenance costs. Valves that consume a large fraction of the total pressure drop for the system, or are excessively throttled, can be opportunities for energy savings. Pressure drops or head losses in liquid pumping systems increase the energy requirements of these systems. Pressure drops are caused by resistance or friction in piping and in bends, elbows, or joints, as well as by throttling across the control valves. The power required to overcome a pressure drop is proportional to both the fluid flow rate (given in gallons per minute [gpm]) and the magnitude of the pressure drop (expressed in ft of head). The friction loss and pressure drop caused by fluids flowing through valves and fittings depend on the size and type of pipe and fittings used, the roughness of interior surfaces, and the fluid flow rate and viscosity.

Case study 1 –

In the Hydrocracker Unit of a petroleum refinery the reflux pump generally pumps Naptha with a specific gravity of 0.7. The motor KW rating was 190 kw. Due to some adverse situation in the process, sometimes water also comes in the stream resulting in increase of specific gravity of 0.9.As power P=(wQH/3960) where w is specific weight, Q is discharge and H is the manometric head As a result the motor draws more current. If it goes unnoticed the energy will be wasted.

Case study 2 -

The pump develops the same head in meters of liquid independent of specific gravity. The pump delivers the same quantity by volume independent of specific gravity but the quantity by weight will be proportional to specific gravity. The input power and the output power for a given volumetric discharge vary in proportion to density.

In the Vacuum Distillation Unit of a petroleum refinery the column bottom pump pumps Short residue with a specific gravity of approximately 0.9. Most of the refinery are configured for different type of crudes, like high sulphur and low sulphur. During processing of high sulphur crude the specific gravity and viscosity of the short residue increases. This results in more power consumption.

Case study 3 -

In systems with highly variable loads, pumps that are sized to handle the largest loads may be oversized for normal operating loads. In these cases, the use of multiple pumps, multi-speed motors, or variable speed drives often improves system performance over the range of operating conditions.

To handle wide variations in flow, multiple pumps are often used in a parallel configuration. This arrangement allows pumps to be energized and de-energized to meet system needs. One way to arrange pumps in parallel is to use two or more pumps of the same type. Alternatively, pumps with different flow rates can be installed in parallel and configured such that the small pump operates during normal conditions while the larger pump operates during periods of high demand.

In the Crude unit of a petroleum refinery the main crude pumps are installed in parallel. Some times the crude unit thru put depends upon the secondary unit thruput. If there is any disturbance in the secondary units, then the load in the Crude unit has to brought down. In this case multiple pumps are used in a parallel configuration. This arrangement allows saving energy.

Case study 4 -

The dimension of the pump's flow path may change over time due to scale, rust or sedimentation which usually reduces the size of the pathways. The latter is particularly apt to occur in pumps operating intermittently.

The sulphur pit pump in a Sulphur recovery unit operates intermittently. It pumps liquid sulphur in a jacketed pipe. If the steam in the jacket is not circulated properly or due to ageing, some sulphur particles get deposited in the line. This reduces the flow path size. An increase in pipe roughness increases the friction loss and usually reduces the efficiency. Because of these reasons the pump started drawing more power and tripped more frequently. So to avoid the deposition of sulphur in the line steam was injected so as to clean the sulphur line.

5. Conclusion

The following are different ways to conserve the Energy in Pumping System:

- When actual operating conditions are widely different (head or flow variation by more than 25 to 30%) than design conditions, replacements by appropriately sized pumps must be considered.
- Operating multiple pumps in either series or parallel as per requirement.
- Reduction in number of pumps (when System Pressure requirement, Head and Flow requirement is less).
- By improving the piping design to reduce Frictional Head Loss
- By reducing number of bends and valves in the piping system.
- By avoiding throttling process to reduce the flow requirement.
- By Trimming or replacing the Impellers when capacity requirement is low.
- By using Variable Speed Drives

6. References

[1] "Trouble shooting Process Operations", 3rd Edition 1991, Norman P.Lieberman, PennWell Books
[2] "Centrifugal pumps operation at off-design conditions", Chemical Processing April,May, June 1987, Igor J. Karassik
[3] "Understanding NPSH for Pumps", Technical Publishing Co. 1975, Travis F. Glover
[4] "Centrifugal Pumps for General Refinery Services", Refining Department, API Standard 610, 6th Edition, January 1981
[5] "Controlling Centrifugal Pumps", Hydrocarbon Processing, July 1995, Walter Driedger

Permissions

The contributors of this book come from diverse backgrounds, making this book a truly international effort. This book will bring forth new frontiers with its revolutionizing research information and detailed analysis of the nascent developments around the world.

We would like to thank Prof. Papantonis Dimitris, for lending his expertise to make the book truly unique. He has played a crucial role in the development of this book. Without his invaluable contribution this book wouldn't have been possible. He has made vital efforts to compile up to date information on the varied aspects of this subject to make this book a valuable addition to the collection of many professionals and students.

This book was conceptualized with the vision of imparting up-to-date information and advanced data in this field. To ensure the same, a matchless editorial board was set up. Every individual on the board went through rigorous rounds of assessment to prove their worth. After which they invested a large part of their time researching and compiling the most relevant data for our readers. Conferences and sessions were held from time to time between the editorial board and the contributing authors to present the data in the most comprehensible form. The editorial team has worked tirelessly to provide valuable and valid information to help people across the globe.

Every chapter published in this book has been scrutinized by our experts. Their significance has been extensively debated. The topics covered herein carry significant findings which will fuel the growth of the discipline. They may even be implemented as practical applications or may be referred to as a beginning point for another development. Chapters in this book were first published by InTech; hereby published with permission under the Creative Commons Attribution License or equivalent.

The editorial board has been involved in producing this book since its inception. They have spent rigorous hours researching and exploring the diverse topics which have resulted in the successful publishing of this book. They have passed on their knowledge of decades through this book. To expedite this challenging task, the publisher supported the team at every step. A small team of assistant editors was also appointed to further simplify the editing procedure and attain best results for the readers.

Our editorial team has been hand-picked from every corner of the world. Their multi-ethnicity adds dynamic inputs to the discussions which result in innovative outcomes. These outcomes are then further discussed with the researchers and contributors who give their valuable feedback and opinion regarding the same. The feedback is then collaborated with the researches and they are edited in a comprehensive manner to aid the understanding of the subject.

Apart from the editorial board, the designing team has also invested a significant amount of their time in understanding the subject and creating the most relevant covers. They scrutinized every image to scout for the most suitable representation of the subject and create an appropriate cover for the book.

The publishing team has been involved in this book since its early stages. They were actively engaged in every process, be it collecting the data, connecting with the contributors or procuring relevant information. The team has been an ardent support to the editorial, designing and production team. Their endless efforts to recruit the best for this project, has resulted in the accomplishment of this book. They are a veteran in the field of academics and their pool of knowledge is as vast as their experience in printing. Their expertise and guidance has proved useful at every step. Their uncompromising quality standards have made this book an exceptional effort. Their encouragement from time to time has been an inspiration for everyone.

The publisher and the editorial board hope that this book will prove to be a valuable piece of knowledge for researchers, students, practitioners and scholars across the globe.

List of Contributors

Xiaomei Guo, Zuchao Zhu, Baoling Cui and Yi Li
The Laboratory of Fluid Transmission and Application, Zhejiang Science Technology University, China

Milos Teodor
"Politehnica" University of Timisoara, Romania

Parasuram P. Harihara
Corning Incorporated, USA

Alexander G. Parlos
Texas A&M University, USA

Cristian Patrascioiu
Petroleum Gas University of Ploiesti, Romania

Trinath Sahoo
M/S Indian Oil Corporation ltd., Mathura Refinery, India

Printed in the USA
CPSIA information can be obtained
at www.ICGtesting.com
JSHW011325221024
72173JS00003B/65